INERA Workshop 2016: Membrane and Liquid Crystal Nanostructures (MELINA 2016)

Journal of Physics: Conference Series Volume 780

Varna, Bulgaria
3 – 6 September 2016

Editors:

Julia Genova
Diana Nesheva
Isak Bivas
Yordan Marinov
Victoria Vitkova
Georgi Hadjichristov

ISBN: 978-1-5108-3602-0
ISSN: 1742-6588

Printed from e-media with permission by:

Curran Associates, Inc.
57 Morehouse Lane
Red Hook, NY 12571

CURRAN ASSOCIATES INC. proceedings.com

Some format issues inherent in the e-media version may also appear in this print version.

Printed by Curran Associates, Inc. (2017)

For permission requests, please contact the Institute of Physics
at the address below.

Institute of Physics
Dirac House, Temple Back
Bristol BS1 6BE UK

Phone: 44 1 17 929 7481
Fax: 44 1 17 920 0979

techtracking@iop.org

Additional copies of this publication are available from:

Curran Associates, Inc.
57 Morehouse Lane
Red Hook, NY 12571 USA
Phone: 845-758-0400
Fax: 845-758-2633
Email: curran@proceedings.com
Web: www.proceedings.com

Table of contents

Volume 780

INERA Workshop 2016: Membrane and Liquid Crystal Nanostructures (MELINA 2016)

3–6 September 2016, Varna, Bulgaria

Accepted papers received: 1 December 2016
Published online: 6 February 2017

Preface

Papers

Oral presentations

Posters

PREFACE

We are pleased to introduce the Proceedings of the INERA Workshop "Membrane and Liquid Crystal Nanostructures" (MELINA 2016). This meeting was held from 3th to 6th of September 2016 in Varna (Bulgaria) and was organized by the Institute of Solid State Physics, Bulgarian Academy of Sciences (ISSP-BAS) in the frames of the Project INERA "Research and Innovation Capacity Strengthening of ISSP-BAS in Multifunctional Nanostructures" REGPOT-2012-2013-1 NMP (http://www.inera.org). The workshop venue was the "Saints Constantine and Helena" Resort - a beautiful complex on the Bulgarian Black Sea coast within a landscaped park 10 km north of downtown Varna, close to the Euxinograd royal summer palace, park and winery.

The aim of this workshop was to bring together researchers with interests in interdisciplinary areas in order to report the recent results of their scientific work, to discuss the state-of-the-art & perspectives and communicate new ideas. Several experimental & theoretical streams in the field of nanostructured liquid crystal systems and nanocomposites; soft and living matter physics; membrane biophysics; nanostructures in polymer and lipid membranes; characterization of soft nanomaterials were merged and directed towards a constructive discussion & solutions of a wide variety of actual problems in these important scientific areas. Participants from 9 countries presented 20 oral and 15 poster presentations in 5 topical areas.

INERA Workshop MELINA 2016 provided very good opportunities for direct contacts, interesting discussions and interactive exchange of ideas between the participants. It also became a meeting point where INERA partners could plan and organize their future efficient collaborations and joint research projects of common interest.

The editors would like to thank all authors for their high-quality contributions. The papers submitted for publication in the Proceedings were refereed according to the publishing standards of the Journal of Physics: Conference Series. The proceedings of conferences and workshops organized in the frames of INERA Project are regularly published by the Journal of Physics: Conference Series. The Editorial Committee members are very grateful to the journal's staff for providing us this opportunity, as well as for the continuous support for the publication of these Proceedings.

Managing Editor: Assoc. Prof. Dr. Julia Genova
Co-editors:
Prof. DSc. Diana Nesheva
Prof. DSc. Isak Bivas
Assoc. Prof. Dr. Yordan Marinov
Assoc. Prof. Dr. Victoria Vitkova
Assoc. Prof. Dr. Georgi Hadjichristov

IOP Conf. Series: Journal of Physics: Conf. Series **780** (2017) 011002 doi:10.1088/1742-6596/780/1/011002

INERA Workshop 2016

Membrane and Liquid Crystal Nanostructures

(MELINA-2016)

3-6 September, 2016, Varna, Bulgaria

Topics:

- Nanostructured liquid crystal systems and nanocomposites
- Soft and living matter physics
- Membrane biophysics
- Nanostructures in polymer and lipid membranes
- Characterization of soft nanomaterials

Chairman: Academician Alexander G. Petrov – ISSP-BAS, INERA coordinator

Vice-Chairman: Prof. DSc. Kiril Blagoev – ISSP-BAS

Scientific Secretary: Prof. DSc. Minko Petrov – ISSP-BAS

Local organizing committee:

Prof. DSc. Hassan Chamati – Director ISSP-BAS
Prof. DSc. Kiril Blagoev – INERA and ISSP-BAS Financial Director
Assoc. Prof. Dr. Emil Vlakhov – INERA Manager ISSP-BAS
Prof. DSc. Kostadinka Gesheva – INERA WP Leader
Assoc. Prof. Dr. Yordan Marinov – INERA WG4 Leader
Assoc. Prof. Dr. Victoria Vitkova
Assoc. Prof. Dr. Julia Genova
Assoc. Prof. Dr. Ekaterina Radeva
Assoc. Prof. Dr. Georgi Hadjichristov
Assist. Prof. Lidia Popova

Editorial committee:

Assoc. Prof. Dr. Julia Genova (Managing Editor)
Prof. DSc. Diana Nesheva
Prof. DSc. Isak Bivas
Assoc. Prof. Dr. Yordan Marinov
Assoc. Prof. Dr. Victoria Vitkova
Assoc. Prof. Dr. Georgi Hadjichristov

IOP Conf. Series: Journal of Physics: Conf. Series **780** (2017) 011003 doi:10.1088/1742-6596/780/1/011003

Peer review statement

All papers published in this volume of *Journal of Physics: Conference Series* have been peer reviewed through processes administered by the proceedings Editors. Reviews were conducted by expert referees to the professional and scientific standards expected of a proceedings journal published by IOP Publishing.

Mechano-capacitive properties of polarized membranes and the application to conductance measurements of lipid membrane patches

Karis A Zecchi, Lars D Mosgaard and Thomas Heimburg*

Niels Bohr Institute, University of Copenhagen, Blegdamsvej 17, 2100 Copenhagen Ø, Denmark

E-mail: theimbu@nbi.ku.dk

Abstract. Biological membranes are capacitors that can be charged by applying an electric field across the membrane. The charges on the capacitor exert a force on the membrane that leads to electrostriction, i.e., a thinning of the membrane. This effect is especially strong close to chain melting transitions. A consequence is voltage induced pore formation in the lipid membrane. Since the force is quadratic in voltage, negative and positive voltages have an identical influence on the physics of symmetric membranes. This is not the case for a membrane with an asymmetry leading to a permanent electric polarization. Positive and negative voltages of identical magnitude lead to different physical properties. Such an asymmetry can originate from a lipid composition that is different on the two monolayers of the membrane, or from membrane curvature. The latter effect is called flexoelectricity. It was investigated in detail by A.G. Petrov in the recent decades. As a consequence of permanent polarization, the membrane capacitor is discharged at a voltage different from zero. This leads to interesting electrical phenomena such as outward or inward rectification of membrane permeability. The changes in current-voltage relationships are consistent with the known magnitude of the flexoelectric effect.

1. Introduction

The major components of biological membranes are lipids and proteins. Lipids are amphiphilic molecules with hydrophobic chains and polar head groups. When exposed to water, they spontaneously form membranes, which are thin layers with a thickness of approximately 5 nm. Biological membranes resemble artificial membranes but have proteins embedded into them. They are surrounded by electrolytic buffers containing ions that can associate with the surface. Therefore, membranes are excellent capacitors. Biological membranes usually are asymmetric. For instance, they frequently possess a larger concentration of charged lipids on the inner layer of a biological membrane. Asymmetric proteins can contribute to such an electrical asymmetry, and so can asymmetric ion distributions due to the interaction of ions with lipids. Such effects lead to a net polarization of the membrane. A further effect leading to polarization is membrane curvature. This effect is called flexoelectricity and was described in much detail by A. G. Petrov (e.g.,[1]).

Many functions of biological membranes are associated with electrical phenomena. For instance, the nerve pulse consists of a propagating pulse in membrane voltage and polarization [2]. Traditionally, the electrical phenomena are attributed to the function of certain proteins called ion channels [3]. However, it has been shown that the lipid layer alone can support electrical pulses [4, 5, 6, 7]. Further, in electrophysiological models it is usually assumed that the membrane capacitance is constant, and that the membrane is an electrical insulator. Both assumptions are incorrect. Biological membranes possess melting transitions close to physiological temperatures, where the order of both the lipid chains and the lateral molecular arrangement decreases. In these transitions, both membrane area and thickness changes, giving rise to changes in capacitance by at least 50% [8]. Further, membranes are in fact quite permeable in this temperature regime, and one finds electrical conductance traces that resemble those of ion channel proteins (e.g., [9, 10, 11]). As a consequence, the membrane can display interesting nonlinear conductance phenomena that are similar but not related to the properties ascribed to proteins [12].

In this report we show the temperature-dependent changes in membrane capacitance. Further, we theoretically discuss the emergence of polarization effects in the lipid membrane, that lead to flexoelectricity and a net membrane polarization. These phenomena are used to describe non-linear electrical conductance phenomena that are not symmetric upon reversal of voltage.

2. Materials and Methods

2.1. Chemicals and membranes

1,2-dimyristoyl-*sn*-glycero-3-phosphocholine (DMPC), 1,2-dilauroyl-*sn*-glycero-3-phospho-choline (DLPC),1-palmitoyl-2-oleoyl-*sn*-glycero-3-phosphocholine (POPC), 1-palmitoyl-2-oleoyl-*sn*-glycero-3-phosphoethanolamine (POPE) and cholesterol were purchased from Avanti Polar Lipids (Alabaster/AL, US) and used without further purification. In the experiments shown here two lipid mixtures were used: POPE:POPC=8:2 (mol:mol) for the capacitance measurements and a 10:1=DMPC:DLPC (mol:mol) mixture in the absence and presence of 15% cholesterol (mol/mol of total lipid) for the current-voltage curve measurement.

The 150 mM KCl electrolyte solution used for the BLM experiments was buffered with 2 mM HEPES and 1 mM EDTA (both from Sigma Aldrich, Germany), pH 7.4. For the patch clamp experiments, the electrolyte was made of 150mM KCl, 150 mM NaCl and 50 mM TRIS (Sigma Aldrich, Germany) and the pH was 7.6. All water used in the experiments was purified with a Direct-Q® Water Purification System (Merck Millipore, Germany) and had a resistivity >18.1 MΩ·cm. Electrolyte solutions were filtered through a sterile 0.2 μm filter (Minisart® Sartorius Stedim Biotech, Germany) to get rid of dust particles or impurities (this is particularly important in experiments with glass pipettes of small aperture).

2.2. Black lipid membrane (BLM) setup

The DMPC:DLPC:cholesterol = 77.3:7.7:15 (mol:mol:mol) mixture used for the current-voltage (I-V) measurements and the POPE:POPC = 8:2 (mol:mol) mixture used for the capacitance measurements on the Black Lipid Membrane (BLM) setup were both dissolved in decane to a final concentration of 10 mg/mL. Planar lipid bilayers were formed on a circular aperture in a 25 μm thick Teflon film using a slightly modified version of the painting method introduced by Mueller et al [13]. The original method consists in painting a small volume of the lipid solution on the hole of a Teflon film separating two chambers filled with an electrolyte solution. The membrane is then allowed to thin out for a few minutes until a bilayer is formed which is in equilibrium with a surrounding annulus made of the bulk lipid solution (the so called Plateau-Gibbs border), as shown in figure 1 (left). In this particular example, the hole had a diameter of 120 μm, and the membrane patch a diameter of $\varnothing$=86 μm. More details on the different methods for bilayer formation can be found in [14]. Here we used commercially available horizontal bilayer slides (Ionovation GmbH, Germany) made of two microchambers (filled with approx 150 μL of the same electrolyte solution) separated by an horizontal Teflon film. The upper and lower

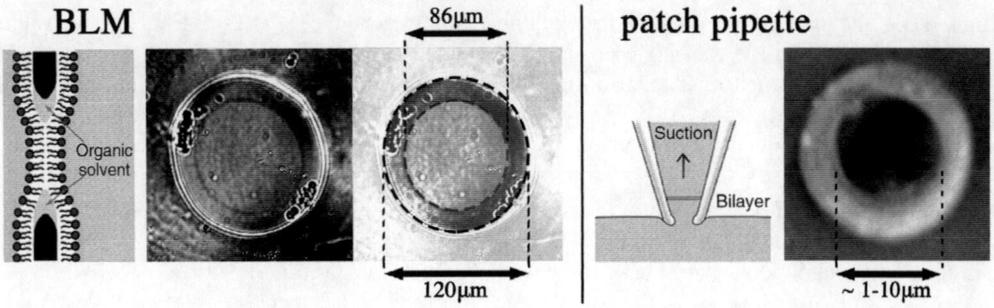

Figure 1. Left: Schematic drawing [14] and image of black lipid membrane. The outer diameter of the hole (~120 µm) and the diameter of the membrane (~86 µm) are highlighted by dashed lines. Right: Schematic drawing of a membrane in a patch pipette, and microscopic image of a patch pipette tip with diameter of ~1- 10 µm diameter.

chambers are connected only through the 120µm aperture in the film and each has an access port for the electrical connections and for buffer perfusion. Once a small droplet ($\simeq 0.2$ µL) of lipid solution is placed in the upper chamber close to aperture, a bilayer is formed automatically by a microfluidic perfusion system (Ionovation Explorer, Ionovation GmbH, Germany). The membrane formation was monitored with capacitance measurements and was automatically repeated until the membrane capacitance was stable above a minimum threshold value of 40 pF. The bilayer slide was placed on the work stage of a in inverted microscope (IX70, Olympus, Japan) which allowed for optical monitoring of the bilayer formation.

2.3. Calorimetry

Heat capacity profiles were recorded using a VP-DSC (MicroCal, Northampton/MA, USA - now part of Malvern Instruments Ltd. ,Worcestershire , UK) with a scan rate of 5°/h.

2.4. Patch Clamp Setup

The mixture of 10:1 = DMPC:DLPC (mol:mol) was resuspended in a highly volatile solvent made of 4:1=hexane:ethanol (vol:vol) to a final concentration of 2mM. Synthetic lipid membranes were formed on the tip of a patch-clamp glass pipette using the droplet method introduced by Hanke [15, 14], in which a droplet of the lipid solution is placed on the outer surface of a vertically standing glass pipette filled with the electrolyte solution. The tip of the pipette is in contact with the surface of a beaker containing the same electrolyte solution as the pipette, and as the droplet flows down the glass surface, it seals the pipette tip with a spontaneously formed bilayer. The solvent is the allowed to diffuse out of the bilayer for about 30s before starting the experiments. The whole experimental procedure for the patch clamp setup followed the one used and explained in [16, 17]. A microscopic image of the tip of a patch pipette is shown in figure 1 (right). Typical membrane patches have a diameter between 1 and 10 µm. Thus, they are much small than the BLMs.

3. Theory

3.1. Capacitance

The capacitance of a membrane is given by

$$C_m = \varepsilon_0 \varepsilon \frac{A}{d} , \tag{1}$$

where A is the membrane area, d its thickness, $\varepsilon_0 = 8.8541878 \cdot 10^{-12}$ F/m is the vacuum permittivity, and ε is the dielectric constant (assumed to be 3 for the hydrophobic core of the membranes). In the melting transition of a lipid membrane made of DPPC, the area changes by 24% and the thickness by -16%. Therefore, the calculated capacitance increase is 49% upon scanning through the melting transition [8].

3.2. Capacitance and polarization

Recently, we have investigated polarization effects in membranes [12]. If one considers the membrane as a capacitor, the membrane charge is given by

$$q = A(\varepsilon_0\varepsilon E + P_0) = \varepsilon_0\varepsilon\frac{A}{d}(\Psi + \Psi_0) = C_m(\Psi + \Psi_0)\ , \tag{2}$$

where P_0 is the permanent polarization of the membrane. The electrostatic potential difference across the membrane is $\Psi = E \cdot d$ and the offset potential related to the polarization is $\Psi_0 = (P_0/\varepsilon_0\varepsilon) \cdot d$. The latter quantity describes the membrane polarization that is present in the absence of an external field. The electrostatic free energy density of the membrane is given by [12]

$$g^{el} = -\frac{C_m}{2\,A}\left((\Psi + \Psi_0)^2 - \Psi_0^2\right)\ , \tag{3}$$

which is zero at zero voltage.

3.3. Flexoelectricity

From equation 2 it follows for the curvature dependence of the charge on the membrane capacitor (at constant temperature, membrane area and constant voltage)

$$dq = \left[(\Psi + \Psi_0)\underbrace{\left(\frac{\partial C_m}{\partial c}\right)_\Psi}_{\equiv\alpha} + C_m\underbrace{\left(\frac{\partial \Psi_0}{\partial c}\right)_\Psi}_{\equiv\beta}\right] dc \equiv \left[(\Psi + \Psi_0)\alpha + C_m\beta\right] dc\ , \tag{4}$$

where α and β are constant coefficients. The curvature is given by $c = 1/R$, where R is the radius of curvature. If the capacitance is constant ($\alpha = 0$), the integral of the above yields

$$q(c) = C_m\left(\Psi + \Psi_0(0)\right) + C_m\beta \cdot c\ , \tag{5}$$

where $C_m\left(\Psi + \Psi_0(0)\right)$ is the membrane charge at $c = 0$ (equation 2). In the absence of an applied voltage ($\Psi = 0$), and when the membrane is not polarized in the absence of curvature ($\Psi_0(0) = 0$) we obtain

$$q(c) = C_m\beta \cdot c \qquad \text{or} \qquad \Psi_0(c) = \beta \cdot c\ , \tag{6}$$

making use of $q/C_m = \Psi_0$ for $\Psi = 0$ (equation 2). Thus, the offset potential Ψ_0 is proportional to the curvature. This relation is a special case of the flexoelectric effect described in equation (4). A. G. Petrov discussed this effect in many papers in the 1980s and 1990s [1]. In his nomenclature,

$$\Psi_0(c) = \frac{f}{\varepsilon_0\varepsilon} \cdot c\ . \tag{7}$$

If the membrane is curved along two axes, this has to be replaced by $\Psi_0(c) = (f/\varepsilon_0\varepsilon) \cdot (c_1 + c_2)$. Here, the flexoelectric coefficient $f \equiv \beta \cdot \varepsilon_0\varepsilon$ is introduced. Petrov and collaborators found experimentally that $f = 1 - 4 \cdot 10^{-18}$ [C] for different membranes and mixtures [18]. Let us assume that $f = 1 \cdot 10^{-18}$ C, which leads to $\beta = 3.8 \cdot 10^{-8}$ [V$\cdot$ m] for $\varepsilon = 3$. For a radius of

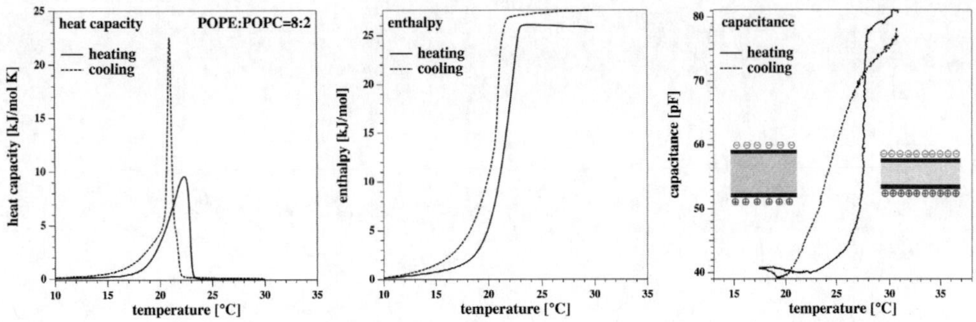

Figure 2. Left: Melting profile of a POPE : POPC = 8:2 membrane. Shown are both heating and cooling scans. Scan rate was 5°/hour. Center: The integral of the heat capacities yields the enthalpy changes. Right: In BLM experiments, one finds significant changes of the membrane capacitance of 89%.

curvature of 100 μm one finds $\Psi_0 = 380$ μV, and for a radius of curvature of 1 μm one obtains $\Psi_0 = 38$ mV. The flexoelectric effect implies that membrane curvature induces a polarization of the membrane. As mentioned, the flexoelectric coefficients were found to be very dependent on the lipid, and it is likely that it is also affected by the phase behavior of lipid mixtures. Generally, the flexoelectric effect seems to be larger in charged membranes.

3.4. Pore formation and conductance

The above concepts can be applied to understand the influence of voltage and membrane polarization on membrane permeability. The permeability of a membrane is related to the formation of small pores in the lipid membrane. The likelihood to form a pore is proportional to the square of the applied electric potential [19, 20, 17]. This assumption is based on the hypothesis that an increase in voltage thins the membrane due to an effect called electrostriction [8] and eventually leads to an electric breakdown linked to pore formation. It has been found that current-voltage (I-V) relations for a chemically symmetric phosphatidylcholine membrane patch can be either a non-linear symmetric [21], or non-linear asymmetric functions of voltage [16, 17, 12].

Blicher et al. [17] proposed that an offset potential, Ψ_0, can explain asymmetric I-V profiles. The free energy difference between an open and a closed pore, ΔG_p, can be expressed by

$$\Delta G_p = \Delta G_{p,0} + \delta\left((\Psi + \Psi_0)^2 - \Psi_0^2\right) , \qquad (8)$$

where $\Delta G_{p,0}$ and δ are coefficients describing the difference in free energy between open and a closed pore in the absence and the presence of externally applied voltage. The equilibrium constant between open and closed pores is given by $K_p = \exp(-\Delta G_p/kT)$, and the likelihood of finding an open pore is given by $P_{open} = K_p/(1 + K_p)$. The I-V relation can be expressed as

$$I = \gamma_p P_{open} \Psi , \qquad (9)$$

where γ_p is the conductance of the membrane when all pores are open. A value for Ψ_0 can be determined by fitting the above set of equations to the experimentally determined I-V profile. The equations have 4 parameters, Ψ_0, $\Delta G_{p,0}$, δ, and γ_p. Only Ψ_0 has an influence on the asymmetry. Therefore, the offset potential is well-defined.

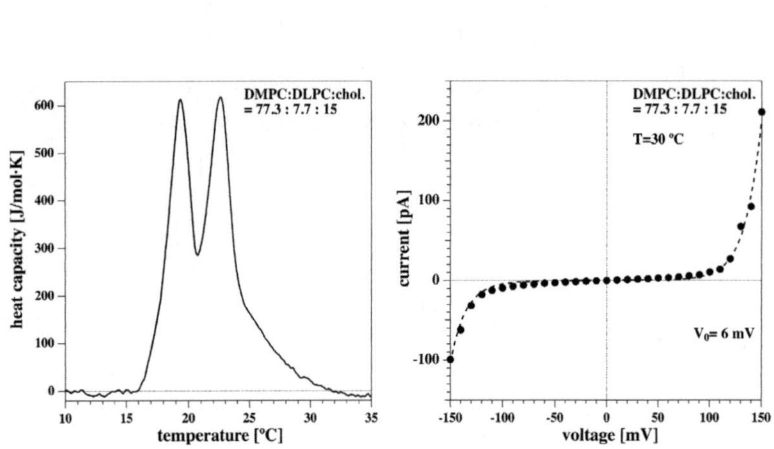

Figure 3. Electrical experiments in BLMs. Left: Melting profile of membranes made from a mixture of DMPC, DLPC and cholesterol (molar ratios 77.3:7.7:15). Right: Current-voltage relationship. It is only slightly asymmetric. The dashed line is a fit using equation 9. The offset voltage determined this way is 6mV.

4. Experimental results

4.1. Capacitance measurements

We first measured the capacitance as a function of temperature in order to demonstrate that it changes in the transition regime, and therefore is not generally constant. For a membrane made of POPE:POPE = 8:2, this effect is shown in figure 2. The left hand panel shows the heat capacity profiles obtained by heating and by cooling. The integral of the heat capacity yields the enthalpy changes, H (figure 2, center). The capacitance, measured in the BLM setup, changes by about 89 ± 6 % upon melting, roughly proportional to the enthalpy changes. It should be noted that the BLM contains decane, which is not present in the calorimetric measurement. This leads to slightly different shapes and values for the transition temperature.

In the following, we use conditions where the membranes are at the upper end of the transition.

4.2. Current-voltage (I-V) relations

In the folling we measure the current through a membrane as a function of voltage at constant temperature. Figure 3 shows a BLM experiment performed on a DMPC:DLPC:cholesterol = 77.3:7.7:10 mixture. In the BLM experiment, the aperture of the hole in the teflon film is usually large - of the order of 100 μm (cf. figure 1, left). The left hand panel of figure 3 shows a heat capacity profile. It displays two heat capacity maxima at 19.4°C and 22.6°C. The profile extends up to 30°C. The right hand panel shows an I-V profile recorded at 30°C, i.e., at the upper end of the transition. We find that is symmetric, meaning that the application of negative and positive voltages yield very similar results. The dashed line indicates a fit to eqs. 8-9. We find a small offset potential, $\Psi_0 \approx 6$ mV. We have recorded many of such profiles. It is a general observation that the I-V profiles turn out being symmetric in BLM experiments (see also [21] for a DOPC:DPPC = 2:1 mixture).

Figure 4 shows a patch pipette experiment performed on a DMPC:DLPC = 90:10 mixture. In pipette experiments, the opening of the pipette tip is usually quite small - of the order of 1 to 10 μm (cf. figure 1, right). The left hand panel of figure 4 shows a heat capacity profile with a maximum at 22.2°C. The right hand panel shows an I-V profile recorded at 24°V, which is at the upper end of the transition range. In the recording technique used by us, the pipette

IOP Conf. Series: Journal of Physics: Conf. Series **780** (2017) 012001 doi:10.1088/1742-6596/780/1/012001

touches the surface of the aqueous buffer. However, one can lower the tip of the pipette and create pressure differences between the two sides of the membrane. In the experiment, we changed the depth of the pipette in the water from 0 mm to 8 mm, corresponding to a pressure change across the membrane of 82 Pa. The I-V profile is quite symmetric in the absence of a pressure difference, while it becomes quite asymmetric upon application of a pressure difference, i.e., application of negative and positive voltages yield very different results. The latter profiles are called 'outward-rectified' meaning that the permeability increases faster with positive as compared to negative voltages. We find an offset potential of $\Psi_0 \approx 0$ mV for the experiment in the absence of a pressure difference, while we find $\Psi_0 \approx 160$ mV at 8mm depth. We often see that changes are transient. This could mean that after a sudden change in pressure material may flow and the tension on the membrane does not stay constant. When repeating this kind of measurement with patch pipettes we nearly always find asymmetric non-linear I-V profiles. It is interesting to note that the I-V profiles shown here for patch pipettes are very similar to those recorded for some membrane proteins [16, 11]. The profiles attributed to proteins as the TRP receptor proteins can equally well be fitted by the simple model leading to eqs. 8-9. Such rectified profiles have been thought to be typical for membrane proteins. However, as we show here, the rectification of I-V profiles easily originates from a polarization of the lipid membrane.

5. Discussion and conclusions

It is obvious that the current-voltage relations of BLMs and of patches on pipettes look different. While the BLM experiments regularly lead to symmetric profiles that become non-linear at higher voltage, they are usually asymmetric on patch pipettes. One can nicely fit the asymmetric profiles with an offset potential, Ψ_0. It is tempting to speculate about the origin of this polarization. One possibility to explain it is related to flexoelectricity. The different diameters of the BLM aperture and the diameter of the patch pipette tip allow for different possible membrane curvatures. In figure 5 we show the scenarios for BLM and patch pipettes. On the left side we show the possible curvatures of a membrane on the opening of a teflon hole. Obviously, a membrane that spans across the hole in a flat manner has a curvature of $c = 0$ [1/m]

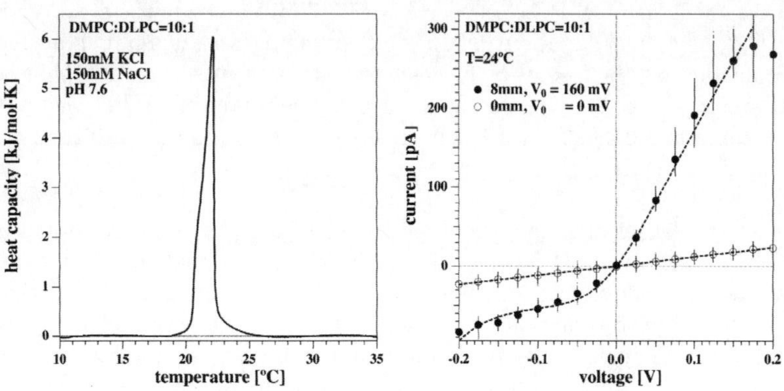

Figure 4. Electrical experiments in patches on a pipette tip. Left: Melting profile of membranes made from a mixture of DMPC and DLPC (molar ratios 90 : 10). The mixture displays one heat capacity maximum at 22.2°C. Right: Current-voltage relationship at two different pressures (difference in water depth is 8mm). If the membrane is at the surface of the aqueous interface, the *I*-*V* curve is symmetric and linear. At 8mm depth, it is asymmetric, and displays an offset potential of 160 mV. The dashed lines are fits using equation 9.

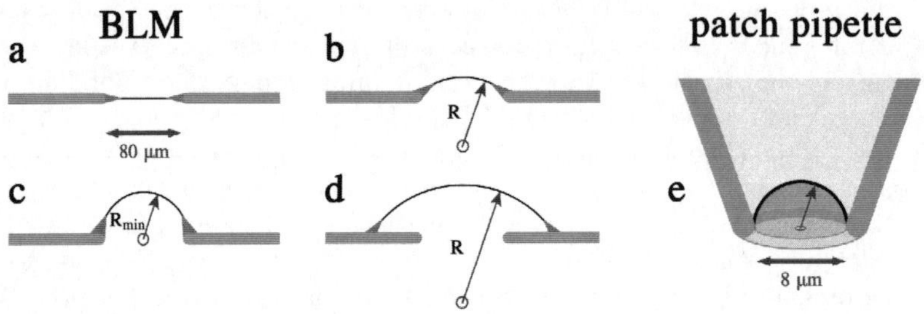

Figure 5. Left: Schematic drawing of a membrane in a BLM measurement. (a) Flat membrane without differences in pressure on the two sides of the membrane. The solvent annulus is shown in in red, an the teflon membrane as a thick grey line. The thin line represents the lipid membrane. (b) In the presence of pressure differences the membranes will bend. (c) The minimum radius of curvature, R_{min} is defined by the dimensions of the hole. Here, it is 40 µm. (d) upon increasing membrane tension, the outer edges of the membranes may slide away from the edges of the hole. Right: The patch pipette has a much smaller diameter. (e) The minimum radius of curvature of the membrane is much smaller, and the curvature much bigger. In the example, the radius of curvature is $\sim$ 4 µm.

(figure 5(a)) and cannot generate an offset potential due to flexoelectricity. However, we often notice in experiments that the membrane bulges out of the plain. The maximum curvature can be expected when the membrane has a radius of curvature similar to the radius of the hole in the membrane (figure 5(c)). All other scenarios lead to smaller curvatures. In figure 5(d) the outer edge of the membrane disconnects from the hole circumference and becomes larger. We often observe this in the experiment. However, in this scenario the curvature becomes smaller. For the present BLM membranes with a diameter of about 80 µm, the minimum radius of curvature is $\sim$ 40 µm. Since the offset potential in the experiment shown in figure 3 is $\Psi_0 = 6mV$, we conclude that for the DMPC:DLPC:cholesterol = 77.3:7.7:15 mixture the flexoelectric coefficient given by $f = \Psi_0 \varepsilon_0 \varepsilon/(c_1 + c_2)$ is about $3 \cdot 10^{-18}$ C assuming a radius of curvature of $\sim$ 40 µm. We have to conclude that the flexoelectric effect expected for BLM membranes is small because curvatures are small. The patch pipette has a much smaller diameter. The offset potential in the experiment shown in figure 4 is $\Psi_0 = 160mV$. If we assume a minimum radius of curvature of $\sim$ 4 µm, we obtain for the DMPC:DLPC = 90:10 mixture a flexoelectric coefficient of $f = 8.5 \cdot 10^{-18}$ C.

The values for the flexoelectric coefficient given above are lower estimates because we assumed a minimum radius of curvature. However, it is interesting to note that these values are reasonably close to the values for the flexoelectric coefficients found by Todorov and collaborators [18]. They found $f = 4.0 \cdot 10^{-18}$ C for bovine brain PS, $f = 2.6 \cdot 10^{-18}$ C for bacterial PE and $f = 1.8 \cdot 10^{-18}$ C for egg PC. The experiments by Todorov et al. were performed on similar membranes but their measurement included a direct measurement of curvature by optical means. Our experiment is an indirect measurement of curvature because we measure a consequence of flexoelectricity rather than the effect itself. However, the closeness of our numbers for f to Todorov's data suggests that flexoelectricity is a very reasonable explanation for the measured effects, which possesses a significant biological relevance. Our studies point at a direct relevance of flexoelectricity to membrane permeability. I-V profiles observed in biological cells that are outward rectified are usually considered to be a unique feature of proteins. We show here that this is not the case.

In order to quantify our experiments in a better manner it would be useful to measure the membrane curvature in a direct optical experiment such as the interferometric microscopic methods used by Todorov et al. [18] and then correlate it with the electrophysiological data. Relating this curvature to current-voltage relations may open a new door to interpret electrophysiological experiments with physical methods.

Acknowledgments
This work was supported by the Villum Foundation (VKR 022130).

6. References

[1] Petrov A G 1999 *The lyotropic state of matter. Molecular physics and living matter physics.* (Amsterdam: Gordon and Breach Science Publishers)
[2] Johnston D and Wu S M S 1995 *Cellular Neurophysiology* (Boston: MIT Press)
[3] Hille B 1992 *Ionic channels of excitable membranes* (Cambridge: Cambridge University Press)
[4] Heimburg T and Jackson A D 2005 *Proc. Natl. Acad. Sci. USA* **102** 9790–9795
[5] Andersen S S L, Jackson A D and Heimburg T 2009 *Progr. Neurobiol.* **88** 104–113
[6] Griesbauer J, Bössinger S, Wixforth A and Schneider M F 2012 *Phys. Rev. Lett.* **108** 198103
[7] Griesbauer J, Bössinger S, Wixforth A and Schneider M F 2012 *Phys. Rev. E* **86** 061909
[8] Heimburg T 2012 *Biophys. J.* **103** 918–929
[9] Blicher A, Wodzinska K, Fidorra M, Winterhalter M and Heimburg T 2009 *Biophys. J.* **96** 4581–4591
[10] Heimburg T 2010 *Biophys. Chem.* **150** 2–22
[11] Mosgaard L D and Heimburg T 2013 *Acc. Chem. Res.* **46** 2966–2976
[12] Mosgaard L D, Zecchi K and Heimburg T 2015 *Soft Matter* **11** 7899–7910
[13] Mueller P, Rudin D O, Tien H T and Wescott W C 1962 *Nature* **194** 979–980
[14] Gutsmann T, Heimburg T, Keyser U, Mahendran K R and Winterhalter M 2015 *Nature Protocols* **10** 188–198
[15] Hanke W, Methfessel C, Wilmsen U and Boheim G 1984 *Bioelectrochem. Bioenerg.* **12** 329–339
[16] Laub K R, Witschas K, Blicher A, Madsen S B, Lückhoff A and Heimburg T 2012 *Biochim. Biophys. Acta* **1818** 1–12
[17] Blicher A and Heimburg T 2013 *PloS ONE* **8** e65707
[18] Todorov A, Petrov A G and Fendler J 1994 *Langmuir* **10** 2344–2350
[19] Winterhalter M and Helfrich W 1987 *Phys. Rev. A* **36** 5874–5876
[20] Glaser R W, Leikin S L, Chernomordik L V, Pastushenko V F and Sokirko A I 1988 *Biochim. Biophys. Acta* **940** 275–287
[21] Wodzinska K, Blicher A and Heimburg T 2009 *Soft Matter* **5** 3319–3330

IOP Conf. Series: Journal of Physics: Conf. Series(2780)7) 012002 doi:10.1088/1742-6596/780/1/0

Boundary potential of lipid bilayers: methods and interpretations

Yu A Ermakov[1] and A M Nesterenko[2]

[1] A N Frumkin Institute of physical chemistry and electrochemistry RAS, Moscow
[2] Biological Faculty, Moscow State University

E-mail: yury.a.ermakov@gmail.com

Abstract. The electric field distribution at the boundaries of cell membrane consists of diffuse part of the electrical double layer and the potential drop over polar area inside the membrane itself. The latter is generally attributed to the dipole effect, which depends on the lipid hydration and phase state. This report focuses on the experimental approaches developed to detect the relation between dipole effects and the bilayer structure, and to study their molecular nature. The total boundary potential (BP) of planar bilayer lipid membranes (BLM) can be controlled by Intramembranous Field Compensation (IFC) method developed in our laboratory. When combined with electrokinetic measurements in liposome suspension it allows detecting the changes of the dipole potential due to adsorption of inorganic cations and charged molecules. Multivalent inorganic cations increase the dipole potential up to 100-150 mV and make the membrane rigid. Most of these observations were simulated by Molecular Dynamics (MD) in order to visualize the relationship of electric field with the different structural factors (lipid structure, water orientation, ion adsorption etc.) responsible for its dipole component. Two principal contributors to BP – water and lipid molecules – create the opposite effects. The negative contribution with respect to the bulk is due to lipid itself and the inorganic cation penetration into the polar area of membrane. The positive contribution is caused by water orientation. Particularly, in the case of lysine adsorption, the contribution of water includes the rearrangement of H-bonds with the lipid phosphate group. This fact explains well the unusual kinetic phenomena registered by IFC in the case of polylysine adsorption at the BLM surface.

The electric field distribution at the boundaries of phospholipid membrane was intensively studied in many different aspects concerning its biological importance. Naturally, this field is sensitive to the ionic composition of the media and becomes essential for any membrane transport of ionized substances. In other hand, many water-soluble drugs primary interact with the outer surface of cells by electrostatic forces. Model lipid systems - liposomes, planar bilayer lipid membranes (BLM) and monolayers – are commonly used to study the electrostatic phenomena and their relation to structure of membrane interfaces. Our presentation focuses on the experimental approaches developed in our laboratory to differ two components of the boundary potential – the electric field in the water solution that corresponds to the diffuse part of the electrical double layer, and the electric potential drop over polar area at the lipid/water interface [1]. The first one, named here as the surface potential, φ_s, is related to screening of surface charge by ionic media as it well defined by the classic Gouy-Chapman model. To describe the ion equilibrium at the membrane surface the classic model was supplemented with Langmuir-type isotherms. We do not present details of this classic theory, known as Gouy-Chapman-Stern (GCS), which is described in many original papers and reviews [2–4]. The second component of the boundary potential (BP), φ_b, is generally attributed to dipoles oriented across the membrane interface and named as a dipole potential, φ_d. This potential is determined by charged groups of lipids, adsorbed substances, and by inorganic ions placed under Helmholtz plane, which corresponds to the outer membrane surface. Naturally, the adsorption of cations shifts the surface potential to positive values but some of them may decrease the dipole potential by their effect on the lipid packing. In any of above-mentioned cases, the adjacent water molecules make the major contribution to this potential. Several examples of this presentation are aimed to illustrate these situations with the aid of methods developed in our laboratory and applied to planar BLM.

Two principal components of the electric field at the membrane/water interfaces are qualitatively illustrated in figure 1.

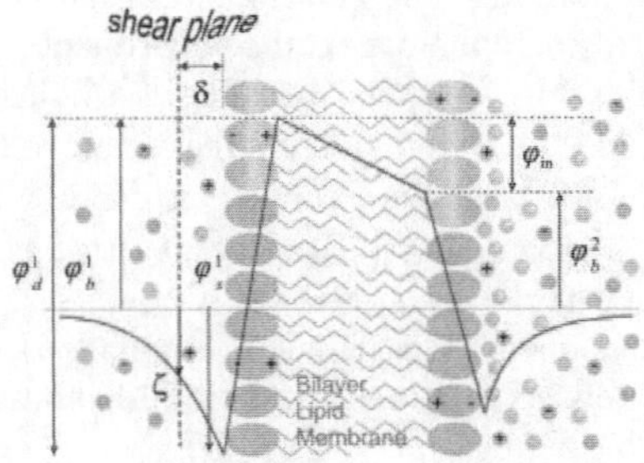

Figure 1. Electric field distribution (red line), the surface (φ_s) and dipole (φ_d) components of boundary potentials (φ_b) at the lipid membrane interfaces. Dotted line illustrate the position of shear plane in the electrokinetic measurements.

The electric field across the membrane is shown in figure 1 by red line. We assume that the membrane has negatively charged lipids and separates a different ionic media at both BLM sides. The electric potential has the same value in the bulk according to short circuit condition. The negative surface potentials (φ_s^1 and φ_s^2) drop at both sides up to zero in the bulk; the dipole components (φ_d^1 and φ_d^2) are directed to high but different positive value inside the membrane. The latter fact follows from many experimental works (i.e. [5−7]). Finally, trans-membrane potential is defined by difference between boundary potentials, which become equal to potential drop over the hydrophobic area of the membrane (intramembranous field) in the short circuit condition:

$$\varphi_{in} = \varphi_b^1 - \varphi_b^2 \tag{1}$$

Some principal points are essential to note. The sign and absolute value of electric potential at the membrane surface is available in electrokinetic measurements at liposome suspension. Data of these measurements shown below were collected from dynamic light scattering experiments. The electrokinetic mobility of liposomes in the typical conditions is related to the electric potential (zeta-potential) at some distance, δ, from the surface in so called "shear plane". The alternative electrostatic methods are required to evaluate this parameter. One of them is illustrated below (figure 4).

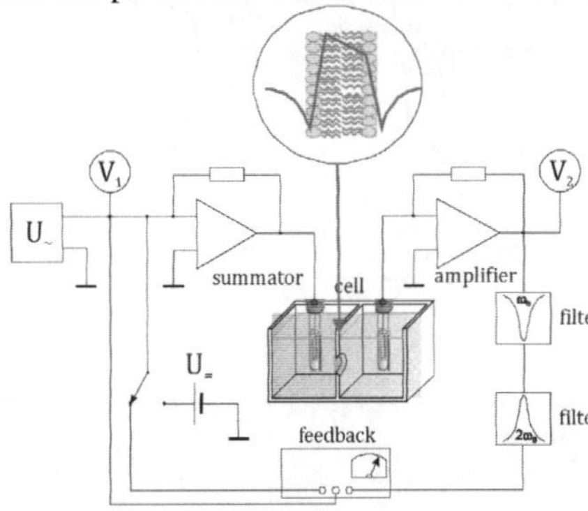

Figure 2. Block-diagram of the setup to measure the difference between the boundary potentials of planar BLM by intramembranous field compensation (IFC) with the help of second harmonic signal caused by membrane electrostriction [1].

Block diagram in figure 2 illustrates the general experimental setup with planar BLM formed in the small hole (about 1 mm in diameter) and some electronics to detect higher harmonics signal of the circuit caused by electrostriction effect [1]. The real setup combines analog devices with their digital versions controlled by computer. It applies a combination of AC and DC voltages to the membrane and register high harmonic signals caused by BLM electrostriction property. In some experiments (figure 3), electric signals in the system registered during the variation of medium composition at one side of the BLM by permanent perfusion the proper cell compartment by peristaltic pump (not shown). Generally, the external DC voltage, V, is applied between the bulks at both membrane sides and compensates the potential drop ϕ_{in} (1). As it follows from the simple analysis of the electric circuit, the capacitive current of "elastic" BLM have the second harmonic with amplitude proportional to difference (ϕ_{in}-V) and the third harmonic – to BLM transversal elasticity. This method, named Intramembrane Field Compensation (IFC), was intensively used in combination with electrokinetic measurements in liposome suspension as a principal instrument to study the electrostatic effects attributed to dipole potential. A few examples listed below demonstrate some advantages of this approach.

Several experiments shown in figure 3 were performed with liposomes and BLM from phosphatidylserine (PS) in solutions of the same concentration (50 mM) but with different monovalent cations – Li^+ and K^+. Left panel shows surface potential deduced from zeta potential due to electrokinetic measurements at different pH. Theoretical curves correspond to Gouy-Chapman model and Langmuir isotherm with parameters – dotted line and solid line. Experiment demonstrates the flat region for totally ionized charged PS with a difference explained by a different binding constants of these cations. Note, that adsorption of potassium cations seems more effective and surface potential is about 15 mV more positive than for lithium cations. Right panel of the figure demonstrates the direct comparison of these cations in perfusion experiment with planar BLM. Arrows indicate the moments of turn on (upper) and turn off (down) the pump, which replaces LiCl solution at one BLM side by KCl of the same concentration (50 mM) or replaces them back. If Li^+ is replaced by K^+ the changes of boundary potential is about 35 mV in negative direction in contrast to electrokinetic data. So, we may conclude that the difference between dipole potential at the membrane boundaries in these solutions was equal to 50 mV. The nature of this difference may be related to different position of these cations in respect to Helmholtz plane, to their effect on lipid packing and hydration of headgroups, to water orientation and H-bonds. All these phenomena defined here as dipole effect and well suitable for future analysis by methods of molecular dynamics.

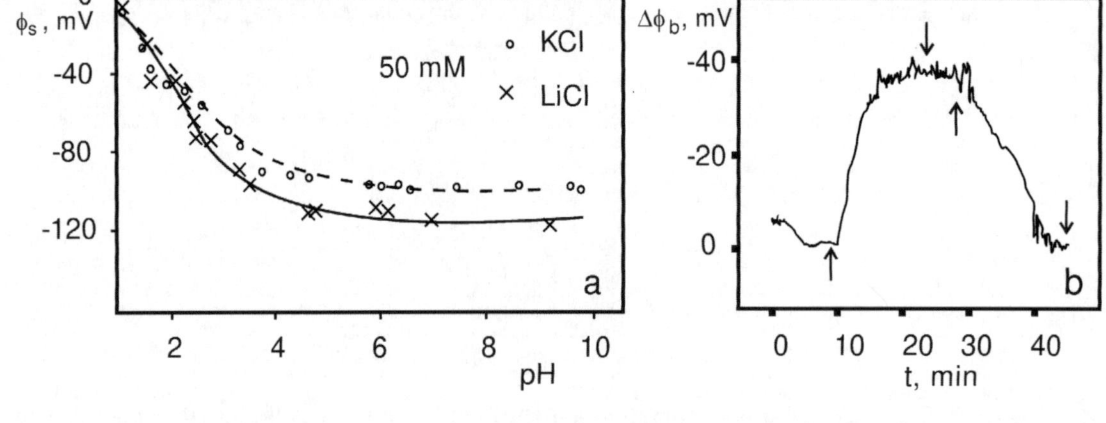

Figure 3. Electrokinetic data (left) for PS liposomes in LiCl (crosses) and KCl (opened points) electrolytes and the boundary potential changes (right) when LiCl is replaced with KCl solution and back by permanent cell perfusion shown by up and down arrows.

The next example (figure 4) demonstrates another type of dipole effect concluded from electrokinetic and IFC experiments with liposomes and planar BLM from dipalmitoyl phosphatidylcholine (DPPC) [8]. Adsorption of divalent cations Be^{2+} at initially uncharged DPPC membranes creates the positive surface charge and potential both at the surface of liposomes and BLM both below and upper the temperature of phase transition. Theoretical curves correspond to Gouy-Chapman model and Langmuir isotherm of Be^{2+} adsorption at PC with a binding constant 400 M^{-1} (45° C, black line) 104 (22° C; dotted line) and shear plane position δ=0.2 nm. Crosses correspond to surface potential recalculated from zeta-potential data (dark points) according Gouy-Chapman theory.

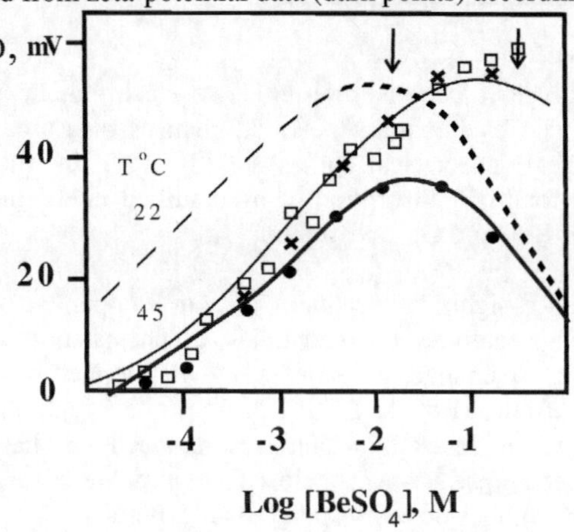

Figure 4. Electrokinetic data for DPPC liposomes (zeta potential) and boundary potential (BP) changes of planar BLM (black and opened points) measured at different $BeSO_4$ concentration. Dotted and solid curves are calculated in the frame of GCS model for temperatures below and behind the point of phase transition. Arrow shows BP changes registered by cooling the experimental cell with BLM [8].

Two conclusions follow from data of figure 4: screening the surface charge by electrolyte of known ionic strength is well described by the model, and the surface potential recalculated from zeta potential with parameter δ=0.2 nm corresponds well to changes of the boundary potential at any salt concentration when the dipole effect is negligible. All measurements presented in figure 4 were done and compared at the temperature 45°C above the phase transition of DPPC because at low temperature planar "solid" BLM became unstable [9]. Parameter δ determine the position of maximum of the curves, which in the case of surface potential is shifted to very high salt concentrations not available in the experiment. It looks natural because GCS model predicts the potential sensitivity to ionic strength of the media (screening effect) increased with a distance (in shear plane) from the surface in water solution. Moreover, the position of maximum of zeta potential depends on the ionic strength only but does not depend on the effectiveness of cation adsorption. Near the same position of maximum appeared for any salts with divalent cations of different affinity to lipid membranes [10].

Dotted line in figure 4 corresponds to electrokinetic data for 22°C with liposomes in 'solid' state, which became more sensitive to Be^{2+} adsorption, and the surface potential is more positive. Unfortunately, at low temperature it is impossible to compare electrokinetic and IFC data directly. Nevertheless, if the experimental cell with "liquid" planar BLM is cooled we may fix the trend of boundary potential changes (shown by arrows at the top of the figure). In contrast to electrokinetic data, the boundary potential became more negative. This fact suggests a strong dipole effect induced by adsorbed Be^{2+} in opposite direction to surface charge and surface component of boundary potential. Recently, this conclusion proved by our MD simulation of the same system [11].

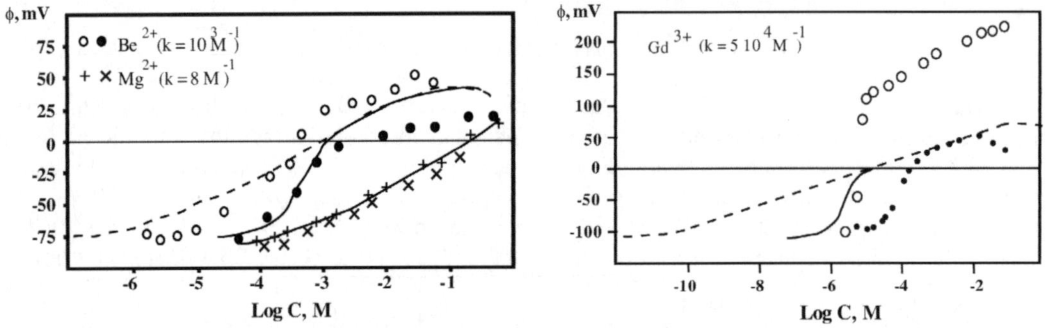

Figure 5. Effect of multivalent cations registered by electrokinetic measurements in liposome suspension (black points) and by boundary potential changes measured by IFC method applied to planar BLM (opened circles). Theoretical curves are calculated by classic GCS model for dotted curves at both panels and for Mg^{2+} adsorption or by modified model included the material balance condition – solid curves for Be^{2+} and Gd^{3+} adsorption [7,8].

Multivalent cations induce significant changes in both component of boundary potential at the membranes composed from negatively charged lipids, as phosphatidylserine in figure 5 [7]. If the cation affinity is not high, these changes are similar and well defined by classic GSC model (Mg^{2+} in figure 5). Cations of high affinity (Be^{2+} and Gd^{3+}) demonstrate a high changes of boundary potential (opened points in both panels in figure 5) which exceeds much the changes of the surface potential recovered from electrokinetic measurements (closed points). Moreover, the slope of experimental curves (dotted lines in figures) does not correspond to GCS predictions. It was observed about 60 mV per decade in the experiment in contrast to the theory with 30 and 20 mV per decade for di- and tri-valent cations, correspondingly. To get the quantitative description of electrokinetic data one has to extend GCS model by the condition of mass balance, i.e. to sum the amount of ions in the bulk, in diffuse and compact parts of the electrical double layer. Note it means that the real bulk concentration of these cations remains unknown because of depletion effect and replaced in figures by their total amount in the experimental cell. According to the modified GCS model cations of high affinity to PS became "potential determining ions" at very low bulk concentration – 10^{-6} or 10^{-10} M for Be^{2+} and Gd^{3+} correspondingly [7,12]. This effect was related recently with the cooperative phenomenon of lipid phase transition of the membrane to "solid" gel state accompanied by lipid-cation cluster formation. This interpretation correlates well with the change of dipole potential induced by high-affinity cation adsorption at the membrane interface. In the examples of figure 5 the difference of boundary and surface potential is clearly observed. According to quantitative analysis of Gd^{3+} adsorption at the lipid membranes of varied composition and pH of the media, the dipole effect induced by these cations is directly related to the presence of PS molecule in their ionized form [7]. Moreover, their interaction initiates the compaction of lipids in the membrane and, as a result, the blocking effect of mechanosensitive channels incorporated into the membrane [13].

Molecular dynamic simulation reveals the opposite input of lipid and water molecules to boundary potential – lipids responsible for high negative and water - for high positive input to BP. The total boundary potential is always positive and is a small difference between two high contributions. MD simulation detects nanoclusters of lipids coordinated with the multivalent cations. Lipid head group hydration is altered significantly in these clusters [11].

IOP Conf. Series: Journal of Physics: Conf. Series **780** (2017) 012002 doi:10.1088/1742-6596/780/1/012002

The similar phenomena are responsible for electrostatic effects induced by adsorption of charged organic molecules – lysine and linear chains of polylysine - to the lipid surface [14]. Two principal facts are important to mention in this aspect (figure 6).

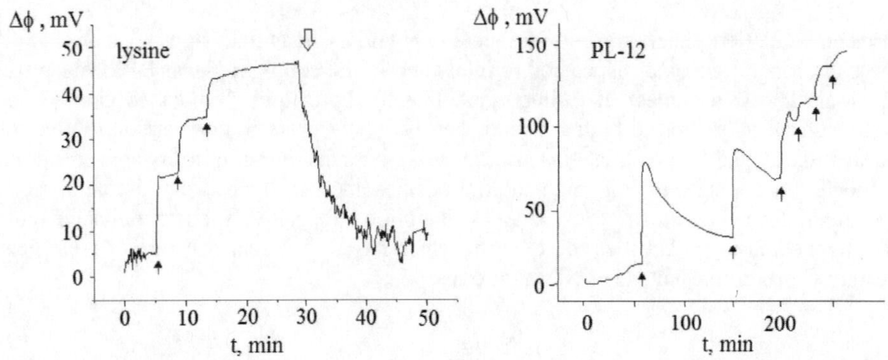

Figure 6. Boundary potential of planar BLM registered by IFC method when lysine added to cell or removed by permanent perfusion at the moment shown by thin and thick arrows in left panel. Right panel illustrate kinetics of boundary potential changes when polylysine stepwise added to one BLM side [14].

Boundary potential at one side of BLM (made from PS) were measured and depicted at figure 6 when lysine or polylysine PL-12 concentration was increased stepwise at moments pointed by black arrows. The solution in the cell was intensively stirred during the experiment and became uniform in a scale of minute or less. The opened arrow in left paned shows the start of cell perfusion by background electrolyte (10 mM KCl). At this step, water solution at both BLM sides returns to their initial symmetry and the magnitude of potential drops to its zero value because the lysine molecules desorbed and were removed from the cell. The same perfusion procedure with polylysines adsorbed at BLM have no effect on boundary potential. This is a general property of many polycations studied in our laboratory, which clearly show their irreversible adsorption at the negatively charged surface of membranes. If a drop of stock solution of polylysines is added to one BLM side (right panel), potential demonstrate two phases of kinetics – fast changes to positive direction and slow return back. The total amplitude of fast steps corresponds to maximal changes of surface potential up to overcharge liposome surface at saturation level detected in electrokinetic measurements. Boundary potential shifted to negative direction (figure 6) indicates the dipole effect. This idea is qualitatively supported by comparison data of two electrostatic methods applied to mono-lysine solutions (figure 7) [14].

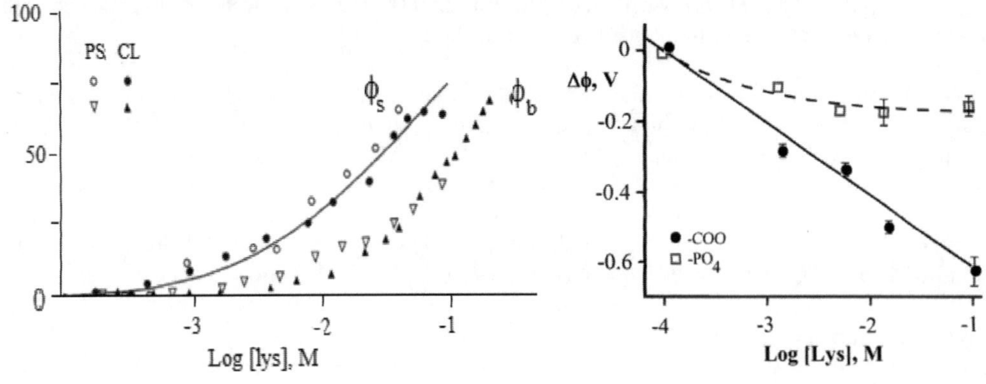

Figure 7. Surface and boundary potentials measured at different lysine concentration in liposome suspension and by IFC method applied to planar BLM composed from PS or Cl (left panel). MD simulation data for dipole effect in DOPS bilayers due to H-bonds to carboxyl and phosphate groups modified by lysine adsorption at the membrane surface (right pane) [15].

Experimental data (left panel) presents surface and boundary potential values measured at different lysine concentration in liposome suspension (circles) and in the cell with planar BLM made from two types of anionic lipids (triangles). It is important, that for both lipid membranes data of boundary potential measurements shifted to high concentration for about order in comparison to electrokinetic data. It means that boundary potential 'insensitive' to lysine adsorption up to its concentration in the cell corresponded to about 50 mV of zeta potential of liposomes. It suggests that dipole component of boundary potential compensate in some extent adsorption of positively charged lysine molecules. Note, no difference observed between these lipids, which have a different structure of polar heads: CL have no carboxyl groups presented in PS head groups.

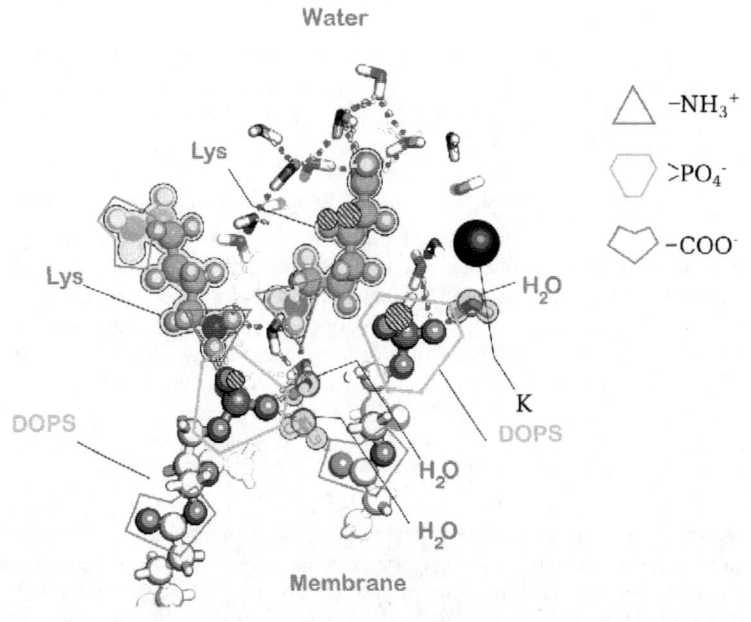

Figure 8. MD simulation data for structural elements in polar area of DOPS membrane in the presence of adsorbed lysine molecules. Dashed lines illustrate bonds directed to water phase, dark sphere is inorganic cation located close to the membrane surface [15].

Digital simulation of PS membrane by methods of molecular dynamics visualizes the structure of polar area in the presence of adsorbed lysine molecules (figure 8) [15]. Dipole effect calculated for this structure reveals a different input of H-bonds between lysine and two negatively charged groups. Comparison with real experiments leads to conclusion that H-bonds with phosphate (but not ones with to carboxyl) shows the saturation effect as it follows from real experiments (figure 7). This conclusion was proved by the similar data observed in the electrostatic experiments with membranes formed from anion lipids PS and CL with the different changed groups.

Conclusions and speculations.
Multivalent cation adsorption induces a strong effect to both BP components and leads to surface overcharging when the cation is bounded to the negatively charged membranes. The effect of

multivalent cations on the lipid lateral interaction is responsible for the dipole potential alterations and for the phase transition of lipids to the solid state. Near the same phenomena are revealed by boundary potential registration during the perfusion procedure. Lysine adsorption was observed at the negatively charged membranes and show the same effect for membranes composed from PS and CL. In both cases, lysine effect to boundary potential became significant at the concentration for about 1.5 order higher than in the electrokinetic measurements. We may propose the compensation between lysine effect on surface and dipole components of BP in the range of several tens of mV. Near the same changes of BP (but to the negative direction) were detected by IFC method in second slow phase of polylysine adsorption at the surface of planar BLM. Probably, these negative changes of PB (figure 6 right panel) are related to the same changes in membrane polar region as is was induced by mono-lysine adsorption. MD simulation of BLM reproduces this lysine effect and demonstrates is relation to H-bonds between water molecules and lipid head groups. Moreover, MD suggests the phosphate groups to be responsible for dipole effects of negative sign in both cases.

Acknowledgements

The work is supported by Russian Foundation for Basic Research (No.16-04-00556a).

References:

[1] Ermakov Yu A, and Sokolov V S 2003 *Planar lipid bilayers (BLMs) and their applications* vol 7 ed H T Tien and A Ottova (Amsterdam: Elsevier) chapter 3 p 109
[2] Cevc G, and Marsh D 1987 *Phospholipid Bilayers. Physical Principles and Models* (New York:Willey-Interscience)
[3] Cevc G 1990 *Biochim. Biophys. Acta* **1031** 311
[4] McLaughlin S 1989 *Annu. Rev. Biophys. Biophys. Chem.* **18** 113
[5] Wang L 2012 *Annu.Rev.Biochem.* **81** 615
[6] Brockman H 1994 *Chemistry & Physics of Lipids* **73** 57
[7] Ermakov Y A, Averbakh A Z, Yusipovich A I and Sukharev S 2001 *Biophys J.* **80** 851
[8] Ermakov Y A, Makhmudova S S and A Z Averbakh 1998 *Colloids and Surfaces A-Physicochemical and Engineering Aspects* **140** 13
[9] Ermakov Y A, Mahmudova S S, Shevchenko E V, and Lobyshev V I 1993 *Biologicheskie membrany* **10** 212
[10] Tatulian S A 1993 *Phospholipid Handbook* ed G Cevc (New York: Marcel Dekker) p 511
[11] Nesterenko A M and Ermakov Y A 2012 *Biochemistry (Moscow) Supplement Series A: Membrane and Cell Biology* **6** 320
[12] Ermakov Yu A 2000 *Colloid Journal* **62** 389
[13] Ermakov Y A, Kamaraju K, Sengupta K, and Sukharev S 2010 *Biophys J.* **98** 1018
[14] Marukovich N, McMurray M, Finogenova O, Nesterenko A, Batishchev O and Ermakov Yu 2013 *Advances in Planar Lipid Bilayers and Liposomes. A Tribute to Marin D. Mitov* vol 17 ed A Iglic and J Genova (Amsterdam: Elsevier) chapter 6 p 139
[15] Marukovich N, Nesterenko A and Ermakov Yu A 2015 *Biochemistry (Moscow) Supplement Series A: Membrane and Cell Biology* **9** 40

Optical nanoscopy characterization of nanofilms

P C Montgomery[1*], P Chapuis[2,3], A Leong-Hoï[1], F Anstotz[1], A Rubin[2], J Baschnagel[2], C Gauthier[2], G Reiter[3] and G B McKenna[4]

[1] Laboratoire des Sciences de l'Ingénieur, de l'Informatique et de l'Imagerie (ICube), Unistra-CNRS, 23 rue du Loess, 67037 Strasbourg, France.
[2] Institut Charles Sadron (ICS), 23 rue du Loess, Strasbourg, France.
[3] Institute of Physics, University of Freiburg, Hermann-Herder-Str. 3, 79104 Freiburg Germany.
[4] Texas Tech University, Dept. of Chemical Eng., Box 43121, Lubbock, TX 79409–3121, USA.

E-mail: paul.montgomery@unistra.fr

Abstract. Glass formation and glassy behavior remains an important field of study in condensed matter physics, with many aspects still little understood. The approach used in this work is to observe the changes in behavior of glass-forming materials at the nanometer scale by exploring the viscoelastic properties of ultrathin free-standing glassy polymer films. An experimental measurement cell based on the nanobubble inflation method is used, consisting of inflating a polymer film suspended over an array of 5 µm diameter holes in a Si wafer. Measuring the deformation as a function of time as the material relaxes is used to determine the creep compliance. Both polystyrene (PS) and poly(vinyl acetate) (PVAc) films of a few tens of nm thickness prepared by spin-coating from solution have been studied. Interference microscopy is used to measure the deformation over several hours, which is challenging at the nanoscale due to mechanical deformations and drift. In this paper we present some of the first solutions developed to allow consistent measurements of film deformation using this novel interference nanoscopy technique. Future work will involve the measurements of creep compliance as a function of different film properties so as to be able to compare the results with theoretical predictions.

1. Introduction

Glass formation and glassy behavior remains an important field of study in condensed matter physics, with many aspects still little understood. An important parameter to consider is the glass transition temperature, T_g, below which glassy materials remain in the glassy state. Above T_g, the materials are in the molten or rubber-like state. It is this property that has allowed significant increases in fuel saving, for example in the aeronautic and automotive industries, through the use of lighter polymer composites to replace metal parts. In bulk materials, T_g remains constant, an important factor for structural design criteria and safe performance. On the contrary, when one of the dimensions is less than about 1 µm, for example in thin films, T_g decreases with thickness [1], which could be problematic for certain industrial applications. One approach to this problem is therefore to observe the changes in behavior of glass-forming materials at the nanometer scale. Some studies state that confinement and/or interfacial physics lead to different behavior for ultra-thin free-stranding polymer films [1, 4]. For example, preparing the film by spin-coating results in an out-of-equilibrium state of the polymer chains [2]. In this work, the viscoelastic properties of ultrathin free-standing glassy polymer films are explored according to the preparation process employed, using an experimental measurement cell based on the nanobubble inflation method [3]. This technique consists of inflating a polymer film suspended over an array of 5 µm diameter holes in a Si wafer held in the cell. Measuring the deformation as a function of time as the material relaxes is used to determine the creep compliance. Using this arrangement, there is no substrate interaction to affect the polymer chains. This is the first part of a longer term project to study the properties of different types of films having a thickness of several tens of nm prepared by different means in order to obtain a deeper insight into the viscoelastic behavior of such ultrathin glassy films.

Both PS and PVAc films of a few tens of nm thickness have been prepared by spin-coating from solution in toluene onto a Si wafer and manually transferring the film onto the perforated substrate using a water bath. Interference microscopy is used to measure the change in deformation over several hours. This technique is one of a family of new unlabeled nanoscopy techniques, consisting of far field optical microscopy that is used for extracting useful structural and physical information from nanostructures [5]. Coherence scanning interferometry (CSI) or white light scanning interferometry (WLSI) as it is also known, using polychromatic illumination, is more commonly used for the measurement of surface roughness and 3D microscopic surface structure [6]. We have adapted this technique to many different applications including the measurement of microelectronic components [7], microsystems [8], biomaterials [9] and colloids [10]. Phase shifting microscopy (PSM), which uses monochromatic or quasi-monochromatic illumination, is used for the measurement of small roughness and surface structures, for example on the surface of optical components [11] and semiconductor materials [12].

Compared with AFM, while interference microscopy is limited in lateral resolution to about $\lambda/2$ (typically about 0.4 µm), it has the advantages of being non-contacting, having a much wider field of view and with a much faster acquisition time. For example, a typical measurement of a static surface using PSM takes only several seconds [13] compared with a typical time of up to 10 minutes or more with AFM. For measurements of periodically moving surfaces, strobing can be used to freeze the surface and PSM used to measure deformations for example in micromirrors up to frequencies of several MHz [14]. In the case of aperiodic motion, real time measurement of changing surfaces has been achieved using a high speed camera and FPGA processing [15]. Although the axial sensitivity of interference microscopy can be better than 1 nm, achieving high accuracy at the nanoscale can be a challenge [16]. Nanoscale measurements of a surface deforming over several hours presents further difficulties due to mechanical deformations and drift. In this paper we present some of the first solutions that we have developed to allow the consistent measurements of film deformation using this novel interference nanoscopy technique.

2. Nano-bubble inflation method

In this section the nanobubble inflation method is explained, showing how measurement of the deformation of nanobubbles over time can be used to determine the creep compliance of the material.

2.1. Experimental measurements of deformation

The polymer film is placed over an array of holes 5 µm in diameter. Measured with AFM, the freestanding film is deformed downwards into the hole by capillary forces (figure 1(a)) [17]. When the array of holes is placed on a pressure cell fed with a supply of air or nitrogen gas in order to inflate the nanobubbles to a set pressure, the nanobubbles inflate. While this deformation can be measured with AFM (figure 1(b)), the technique is not practical for measurements over periods of several hours. The aim of the present work is to use interference microscopy to measure the surface contour of the nanobubble and thus the radius of curvature. If the nanobubble is inflated at room temperature, the deformation is very small even over a period of several hours. The sample needs to be heated up and maintained at constant temperature close to T_g, for example using a Peltier cell. For bulk polymers, T_g

is constant and according to the literature, the values of T_g for nanofilms is less than that for the bulk material [18].

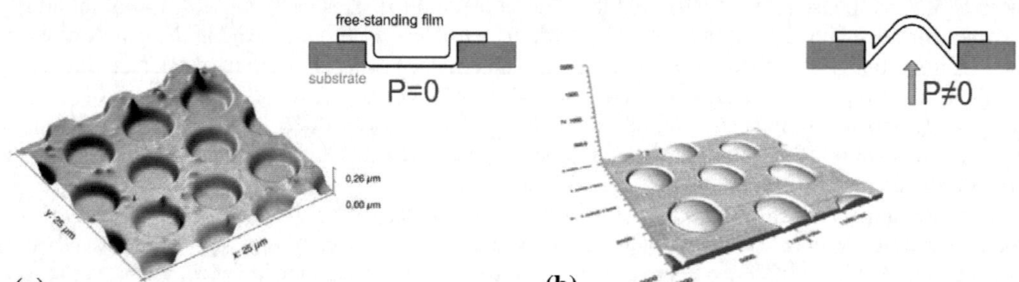

Figure 1. Examples of AFM measurements of films on 5 μm hole grid substrates (a) 90 nm thick PS film with T = RT and P = 0 Pa. Deflection due to capillary forces during the annealing process and (b) inflated 17 nm thick PS film with T = 57 °C and P = 1 psi (≈ 69 kPa) [17].

2.2.Method for estimation of creep compliance

The method used to estimate the creep compliance *D(t)* is the Boltzmann superposition principle for a spherical sample by taking into account the history of the polymer. The first step is to measure the surface contour of the nanobubble at regular intervals over the time period of the experiment (several hours). Then for a contour at a given time, the curvature κ and the radius of curvature *R* are estimated for the average line profile passing through the centre of the circle of the nanobubble by fitting this profile with a second order polynomial *y* and by using the following equations:

$$\kappa = \frac{|y''|}{(1+y'^2)^{3/2}} \qquad R = 1/\kappa$$

This calculation is performed for all the contour measurements of the nanobubble. Then the biaxial components of the stress, σ and the strain, ε are estimated from the radius as a function of time [3]:

$$\sigma = \frac{PR}{2t} \qquad \epsilon = \frac{2Rsin^{-1}(\frac{R_0}{R})}{2R_0} - 1$$

where *P*, *t*, R_0 and *R* are respectively the pressure, the film thickness, the hole radius and the radius of curvature. The Boltzmann superposition principle is then applied to obtain the creep compliance (here *t* denotes time):

$$\epsilon(t) = \int_0^t \sigma(t-\xi)\left[\frac{dD(\xi)}{d\xi}\right]d\xi + D_g\sigma(t)$$

where *ε(t)*, *σ(t)*, *D(t)* and D_g are respectively the time dependent biaxial strain, the time dependent biaxial stress, time dependent biaxial creep compliance and the instantaneous glassy biaxial creep compliance. *D(t)* is then obtained from all the values of *ΔD(t)* estimated from the fits of the experimental data (*σ = f(t)* and *ε = f(t)*).

3. Experimental techniques

In this section, the details of the experimental techniques are given, concerning the preparation of the polymer film samples, the design of the pressure cell and the interference microscopy technique employed to make the deformation measurements.

3.1. Sample preparation

The PS films were prepared by first making a solution of ~1.5% PS/toluene (w/w), with Mw PS = 125 kDa and PDI PS = 1.05. For the PVAc films, these values were ~ 3.0 % PVAc/toluène (w/w), with Mw PVAc = 100 kDa and PDI PVAc = 1.05. To make a film, a clean Si wafer is covered by a thin film of PS or PVAc solution which is then spin-coated at a speed of 2000 rpm. The film is then transferred onto the perforated substrate by floating it in a clean water bath and the samples are allowed to dry overnight in air before finally annealing in an oven for 1 hr at a temperature of T_g + 10°C.

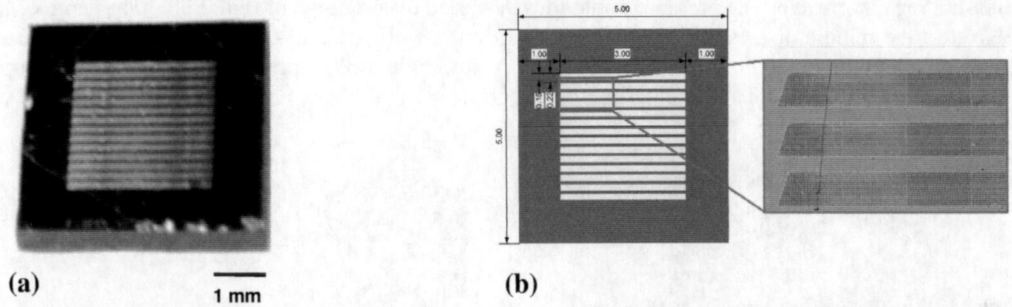

Figure 2. Details of the Si grid substrate (a) photo of Si grid and (b) design of Si substrate (left) and optical microscopy image of small region (right) showing 5 μm holes array with thin PS polymer film (area of greenish tinge); distances in mm.

The perforated grid used is a micro sieve type filter from Aquamarijn Micro Filtration BV (The Netherlands) made by a stencil lithography method [19]. The model employed consists of a very thin (0.9 μm) perforated Si_3N_4 membrane with closely spaced holes 5 μm in diameter in 14 bands (0.15 mm x 3 mm) on a 0.4 mm thick Si support (figure 2). After the deposition of the film, the samples are stored at 30°C ±3°C.

Visual inspection of the polymer film is first performed using optical reflection microscopy. Such an image of the film can be observed in figure 2(b) (far right), the area of greenish tint showing the position of the film placed over the grid of holes.

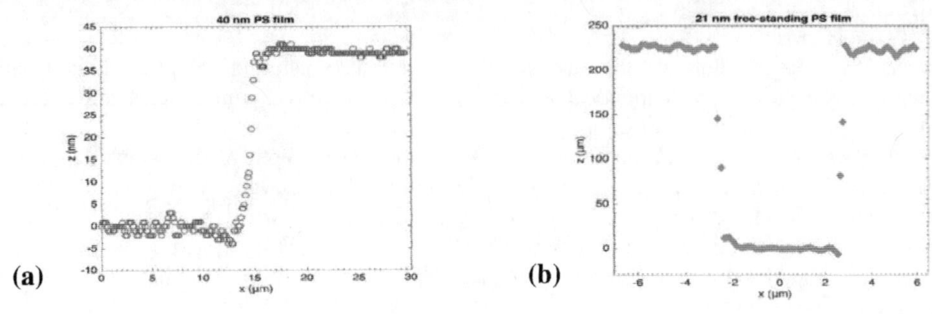

Figure 3. Measurements of PS polymer films using AFM: (a) 40 nm film thickness as measured at edge after placing on grid (b) deflection within hole of a 21 nm thick free standing film due to capillary forces during the annealing process.

Two types of measurements were performed with AFM on a JPK BioMAT Workstation model. First the film thickness was measured near to the edge of the film on a flat part of the perforated grid. An example showing a film thickness of 40 nm of one of the samples is shown in figure 3(a). Then, the shape of a free standing film on one of the holes was measured, the results in figure 3(b) showing a deflection due to capillary forces of about 230 nm into the hole.

3.2. Pressure cell

The pressure cell was designed so as to be able to control the pressure and temperature of the film on the perforated grid (figure 4). The cell was made in aluminium so as to be light enough to be able to be placed on the piezo table (figure 5) for performing fringe scanning. The pressure was provided underneath the film through a hole in the cell by gas from a bottle of nitrogen delivered through a low pressure regulator valve. The pressure could thus be varied over a range of 0-400 kPa. The sample was placed in the middle of a Peltier ring (figure 4) in order to be able to control the temperature from room temperature to 80°C. The temperature of the sample was calibrated using a thermocouple temporarily stuck to the top of an older perforated grid.

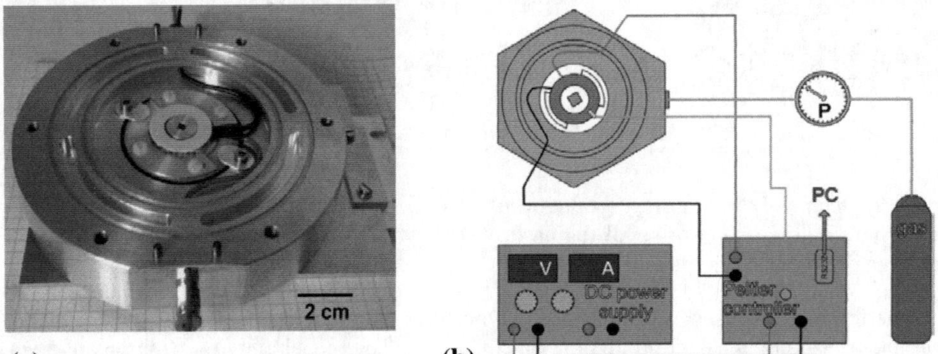

(a) (b)

Figure 4. Details of the pressure cell (a) photo of pressure cell and (b) layout of pressure cell and control of pressure and temperature.

As stated previously, for performing the nanobubble experiments, the temperature of the film was raised to just close to the glass transition temperature of the bulk polymer ($T_g \approx 100$°C for PS).

3.3. Interference microscopy

The deformation measurements of the nanobubbles were made using a modified Leitz-Linnik interference microscope [20] with identical x50 (NA = 0.85) objectives and an incandescent light source. A Pifoc piezoelectric nanopositioner (from PI) controlled in a closed loop with a capacitive position sensor was used for the Z-scanning of the fringes over the depth of the sample (figure 5). Images are acquired with a color CCD camera (Basler avA1000-100gc) with a Giga Ethernet connection. The measurement system is controlled by a PC equipped with an Intel® Xeon® CPU processor (2.40 GHz, 8 Go RAM) under a Windows 7 (64 bits) operating system. The control and analysis software was developed in-house under LabVIEW (version 2014, 64 bits, from National Instruments) combined with the IMAQ Vision module.

The Linnik configuration of the interferometer (figure 5) has the advantages of being able to use high NA objectives, giving a better lateral resolution than the Mirau or Michelson type arrangements [21]. The Linnik interferometer is adjusted first to produce high contrast fringes superimposed on the surface of the film by modifying the path length difference between the two arms to match the coherence planes. The fringes are orientated parallel to the band of holes. The algorithm used to make the phase measurement was a modified version of the 5 phase step algorithm (five steps of $\pi/2$), together with image averaging of 5 frames, which is a good compromise between noise reduction and acquisition time and light (3x3 pixels) median/low pass filtering to reduce the camera noise [20]. Due to the 2π phase discontinuities, the PSM algorithm only allows height measurements up to $\lambda/2$ (< 300 nm at λ = 600 nm). However, the nanobubble growth can achieve a height above 300 nm. This problem can be solved by synchronizing the algorithm with a careful shift control of the piezoelectric table. In this work, a single acquisition with averaging takes ~5 seconds and the whole measurement procedure ~2 minutes, due to manual adjustment of the setup and storing of the results.

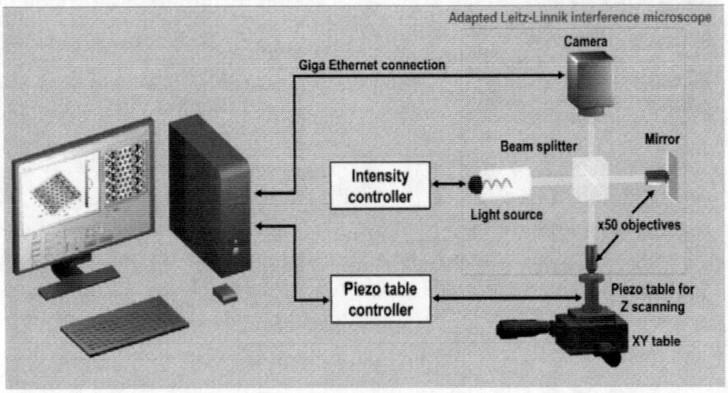

(a)

(b)

Figure 5. Details of the Leitz-Linnik interference microscope system used for nano-3D measurements (a) Schematic layout of system and (b) close-up of the Linnik interferometer and pressure cell.

4. Measurements of the viscoelastic response with time

Making a single measurement of the film surface shape at a given time is fairly straightforward. With a correct disposition of the fringes, the phase shifting technique results in a phase map that can then be converted to height to give a contour map with nm axial sensitivity. Some typical results of a single measurement of the freestanding film are given in figure 6, showing the measurements at two successive steps of post-processing to reduce the noise. The sudden jumps in phase across the image visible in figures 6(a) and (b) are due to the 2π phase discontinuities resulting from the phase shifting technique which results in an axial range limited to $\lambda/2$. In smoothly changing surfaces, the phase can be unwrapped [22] but in the presence of noise or sharp steps, phase unwrapping is not so easy. In these first measurements, due to the presence of noise, by means of a careful choice of the axial position and fringe orientation, direct measurements can be made and the profile read directly (figure 6(c)).

To measure the viscoelastic response of the film, the nanobubble has to be heated to the required stable temperature, inflated to a certain value of pressure and then left to deform over a period of up to several hours. Measurements of the surface contour are then made at regular intervals. The challenge in the present work is to make a series of surface contour measurements over this period of time that can be related to each other meaningfully. When making measurements at the nanoscale, deformations and drift result in profiles that do not seem to be related, since measurements are made on an absolute axial scale.

Several attempts were made to measure the deformation of the nanobubbles over time using the interference technique before arriving at a protocol that enabled fairly consistent results to be obtained of the local deformation. In the following section, some typical results of the measurements are given, together with a discussion of the challenges faced in measuring at the nanoscale and a presentation of some of the first calculations of the viscoelastic response over time.

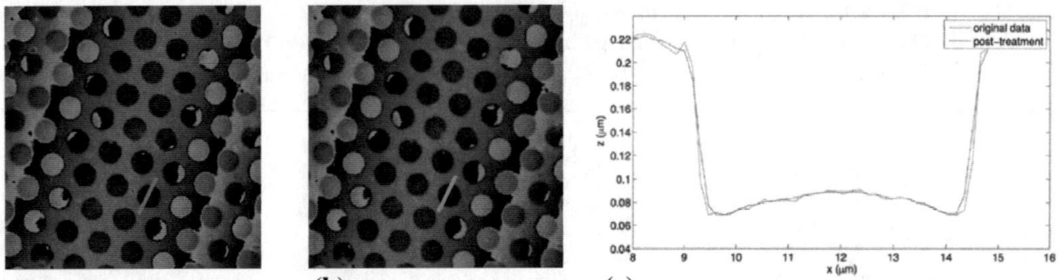

(a) **(b)** **(c)**

Figure 6. Some of the first results of measurements of the surface contour of a freestanding PS polymer film (50 µm x 50 µm region) (a) the phase measurement after median filtering, (b) the phase measurement after median filtering and low pass filtering and (c) profiles of the film over one of the holes after conversion of the phase to height, showing the original data (red line) and after post-processing (blue line).

4.1. Typical measurement results using interferometry

The results in figure 7 show some typical surface contours of an inflated film during one of the experiments measured with the protocol developed. The phase measurements can be observed in figure 7(a), in the presence of phase discontinuities and the associated height contours can be seen in figure 7(b). The correct positions of the nanobubbles is below the surface of the grid, as can be observed in the center near to the line x = 30 µm. To make these nanobubbles visible for the purposes of observation (inset in figure 7(b)), they are "virtually" raised above the level of the grid by using image processing on a small portion of the image to detect the positions of the holes and to apply a mask with an intensity of 0 to lower the surrounding grid.

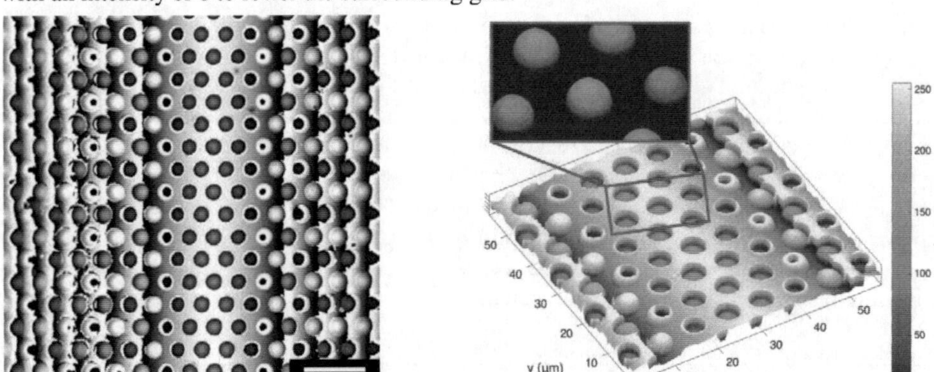

(a) **(b)**

Figure 7. Some of the first results of measurements of the surface contour of an inflated PS polymer film during one of the experiments (a) a 2D image of the measured phase and (b) a 3D image of the height taken from the central region of (a); inset: bubble shapes visualization by hiding the grid.

4.2.Challenges faced in measuring deformation at the nanoscale

As previously noted, measuring the deformation of the nanobubbles at the nanoscale over a period of several hours is challenging. Several of the difficulties met with have already been overcome and some of the solutions developed are now described:

Choice of PSM algorithm for fringe interpretation – Considering that the substrate moves axially over a distance of several μm due to various deformations, a more appropriate fringe interpretation algorithm would have been an envelope detection algorithm in white light. This is commonly used in the measurement of deep surface roughness [21] but with the drawback of having a limited axial sensitivity of about 10-15 nm. Although this value can be improved by combining envelope detection with phase measurement [23, 24], errors at the nm scale can still appear in the case of the presence of steps as in the sample under study. Since the sample was unknown at the beginning of the study, it was therefore decided to use the phase shifting technique, combined with a shift control of the piezoelectric nanopositioner to follow the nanobubble growth, without phase unwrapping so as to stick more closely to the raw data and reduce the number of possible errors. The technique employed was therefore first to align the array of holes symmetrically with the camera field of view and then to optimize the contrast and orientation of the fringes in the Linnik interferometer.

Effects due to the change in temperature – Since some of the samples were heated up to 70°C, it was necessary to wait 1 hr to reach thermal equilibrium, both of the sample cell and the microscope. This was observable by movement of the fringes at the beginning of the heating cycle due to expansion of the different mechanical parts. This movement stabilized after 1 hr. Unknown aspects worth investigating would be the effect of the change in temperature on the sample objective and Linnik interferometer.

Effects due to the pressure – Once the temperature effects were stabilized, several effects on the measurements resulted from the application of the pressure on the film and pressure cell:

1. First, the overall deformation was found to be semi-cylindrical which can be observed in figure 7(a) by the symmetrically arranged phase discontinuities and in figure 7(b) by the curved aspect of the 3D measurement. This is due to the rectangular shape of the array of holes (3 mm x 150 μm) which results in cylindrical deformation under pressure. Taking a profile across the short width of the array of holes therefore gives a curved profile. To avoid the need for correction, the fringes were simply aligned parallel to the longer length of the grids so that the line profiles of the nanobubbles were nearly flat at the center (in the direction of the grid deformation (figure 7(a)).
2. A second effect of the pressure is vertical drift over time of the overall position of the sample along the optical axis, due to the deformation of the substrate and holder. This effect is corrected between measurements by refocusing. Finer adjustment of the fringe position is achieved by using the offset of the piezo nanopositioner, a function which is written into the controlling software.
3. A third effect of the pressure is fast nanobubble growth over the first 10 minutes. Since the present measurement procedure takes ~2 minutes, this means that data acquisition is limited over this important initial period of deformation.

Possible sources of measurement error – After this initial study of the technique for deformation measurement, several other possible sources of error are suggested:

1. Is there a phase on reflection error of the measured wavefront from the sample between the polymer film and the film on the silicon nitride hole array support? Since these materials are both dielectric, in principle this should not be the case, but there could still be some thin film effects. Simulation could help answer this question.
2. Are there edge artefacts of individual nanobubbles influencing the reconstruction of the surface profile of the film? This could be investigated by making different measurements at various tilt angles of the sample.
3. A comparison of the edge measurements of the nanobubble profiles shows the edge positions to be moving upwards (figure 8(a)). Is this due to slipping of the polymer film due to it unsticking at the edge or is it due to the measurement method, or both?

4.3. *Interpretation of the deformation measurements*

The results in figure 8(a) consist of a set of typical measurements, showing the change in the measured profile of a PVAc nanobubble as a function of time over a period of just over 3 hrs. They show a decreasing rate of inflation with time. Using the method described in section 2.2, the creep compliance was then calculated step by step as a function of time by taking into account the stress and strain histories, giving the plot in figure 8(b). What this curve shows is that after a given time, the polymer reaches the rubbery plateau. This means is that at this step, the polymer chains in the film are very entangled while the chain segments are very mobile. Because of this, the whole chain is impeded and the creep compliance becomes and remains constant in time.

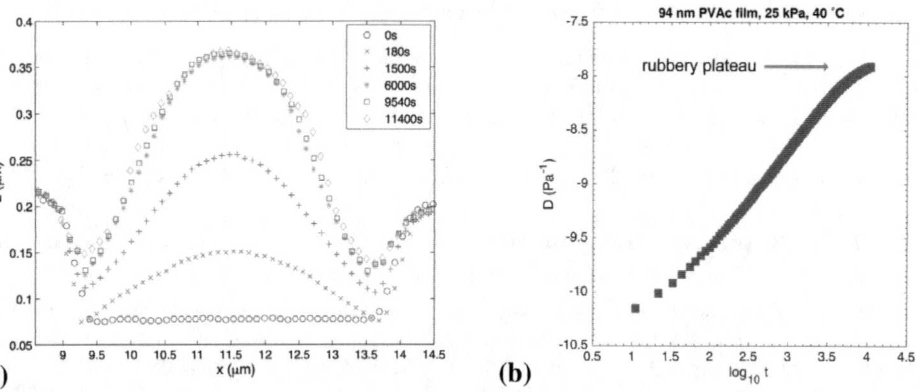

(a) (b)

Figure 8. One set of results of measurements made on a PVAc film nanobubble (a) profiles of the nanobubble at different times and (b) plot of the creep compliance, $\log_{10} D(t)$ versus $\log_{10} t$ (in seconds), from the profile in (a), estimated using the described method.

5. Conclusions

The aim of this work was to investigate the glassy behavior of materials by studying the viscoelastic responses with time of freestanding polymer nanofilms using the Boltzmann superposition principle for spherical samples. The nanobubble inflation method was used in combination with interference microscopy for measuring the surface contour of the nanobubbles over time. To this end, an adapted pressure and temperature cell was specially developed for the microscope.

Measuring nanodeformation over a period of several hours is challenging, due to deformations and drift arising from the use of pressure and heating of the nanofilms. In this paper we have presented some of the first solutions developed to allow consistent measurements of film deformation using a novel interference nanoscopy technique.

Further work is necessary in order to reduce even further the noise and artefacts in the measurement of the surface contour of the nanobubble, such as by improving the optical quality of the Linnik interferometer and by using quasi-monochromatic illumination light. The measurement time could also be reduced well below 2 minutes by automating the acquisition, processing and adjustment procedure.

Future work will then involve the measurements of creep compliance as a function of different film properties, pressures, aging times and Mw. These results will be then compared with existing data measured by AFM [3, 17] with the aim of providing reference data for the viscoelastic behavior of polymer nanofilms, similarly to those existing in the bulk.

6. Acknowledgements
Thanks are given to Damien Flavier for developing the temperature controller, to Stéphane Roques for installing the nitrogen gas pressure system, to Silvia Siegenführ and Asad Jamal for the AFM recommendations and to Sivasurender Chandran for helping with the sample preparation method. Special thank are extended to the IRTG "Soft Matter Science" (GRK 1642/2) for financial support.

References
[1] Forrest J A and Dalnoki-Veress K 2000 *Advances in Colloid and Interface Science* **94** 167
[2] Chowdhury M, Freyberg P, Ziebert F, Yang A C M, Steiner U and Reiter G 2012 *Physical Review Letters* **109** 136102
[3] O'Connell P A and McKenna G B 2005 *Science* **307** 1760
[4] Baschnagel J and Varnik F 2005 *J. Phys.: Condens. Matter* **17** R851
[5] Montgomery P, Leong-Hoi A, Anstotz F, Mitev D, Pramatarova L and Haeberle O 2016 *J. Phys. Conf. Series* **682** 012010
[6] De Groot P 2015 *Advances in Optics and Photonics* **7** 1
[7] Montgomery P, Fillard J P, Castagné M and Montaner D 1992 *Semicond. Sci. and Technol* **7** A237
[8] Montgomery P C, Montaner D, Manzardo O, Flury M and Herzig H P 2004 *Thin Solid Films* **450** 79
[9] Pecheva E, Montgomery P, Montaner D and Pramatarova L 2007 *Langmuir* **23** 3912
[10] Halter E, Montgomery P C, Montaner D, Barillon R, Del Nero M, Galindo C and Georg S 2010 *Applied Surface Science* **256** 6144
[11] Perrot O, Guinvarc'h L, Benhaddou D, Montgomery P C, Rimet R, Boulard B and Jacoboni C 1995 *J. Non-Crys. Solids* **184** 257
[12] Benatmane A, Montgomery P C, Fogarassy E and Zahorsky D 2003 *Applied Surface Science* **208-209** 189
[13] Schmit J, Reed J, Novak E and Gimzewski J K J 2008 *Opt. A: Pure Appl. Opt.* **10** 064001 1
[14] Petitgrand S and Bosseboeuf A 2004 *Journal of Micromechanics and Microengineering* **14** 9
[15] Montgomery P C, Anstotz F, Johnson G and Kiefer R 2008 *Mat. Sci.: Mat. in Elec.* **19** Suppl. 1 194
[16] Montgomery P C, Guellil M, Pfeiffer P, Serio B and Pramatarova L 2014 *J. Phys. Conf. Series* **558** 012005 1
[17] O'Connell P A, Hutcheson S A and McKenna G B 2008 *Journal of Polymer Science: Part B: Polymer Physics* **46** 1952
[18] Dalnoki-Veress K, Forrest J A, Murray C, Gigault C and Dutcher J R 2001 *Physical Review E* **63** 031801
[19] https://www.aquamarijn.nl/showcases/nanostencils/
[20] Montgomery P C, Montaner D and Salzenstein F 2012 *Proc. SPIE* **8430** 843014
[21] De Groot P 2015 *Handbook of Optical Metrology: Principles and Applications* (CRC Press, Boca Raton)
[22] Ghiglia D C and Pritt M D 1998 *Two-Dimensional Phase Unwrapping, Theory, Algorithms, and Software* (John Wiley & Sons, New York)
[23] Harasaki A, Schmit J and Wyant J C 2000 *Applied Optics* **39** 2107
[24] Montgomery P C, Montaner D, Salzenstein F, Serio B and Pfeiffer P 2013 *Proc. SPIE* **8788** 87883G

Novel hydrazones – antioxidant potential and stabilization via polysaccharide particles

N Hristova-Avakumova[1], B Nikolova-Mladenova[2], K Yoncheva[3] and V Hadjimitova*[1]

[1] Department of Medical Physics and Biophysics, Faculty of Medicine, Medical University of Sofia, Bulgaria
[2] Department of Chemistry, Faculty of Pharmacy, Medical University of Sofia, Bulgaria
[3] Department of Pharmaceutical Technology and Biopharmaceutics, Faculty of Pharmacy, Medical University of Sofia, Bulgaria

E-mail: verah@mail.bg

Abstract. In this study, we aimed to: i) determine the impact of three new isonicotinoyl hydrazones derivatives in *in vitro* systems used to investigate free radical processes - radical scavenging approach (ABTS and DPPH) and iron induced peroxidation in lipid containing model systems and ii) evaluate the potential of polysaccharide-based particles to act as protective carriers preserving the antioxidant activity (AOA) of the tested compounds.
The tested compounds revealed excellent antioxidant effectiveness in the ABTS system. In the DPPH radical scavenging assay the compounds exhibited very weak or absence of AOA. The data from the iron induced peroxidation methods disclosed better antioxidant properties of the derivatives in the system containing egg yolk homogenate which is more plausible compared to the lecithin containing one.
The incorporation of a bromine atom on 5th position in salicylaldehyde moiety is associated with diminishment of the radical scavenging activity in the systems containing stable free radicals but its AOA reduction after encapsulation during the storage was only 9.17%. The obtained data indicate that compounds have proven themselves as promising candidates for further evaluation as antioxidant agents. Their encapsulation in chitosan-alginate particles could be a useful approach for improving the stability of their antioxidant properties.

1. Introduction

During the past decades there has been an abundant amount of evidence in the scientific literature proving the connection between oxidants, free radical reactions and the pathogenesis of several incurable progressive illnesses which are among the main causes of death worldwide [1]. This fact has led to the increasing interest in the synthesis and development of novel antioxidants suppressing and preventing free radical cellular damage. The process of discovery of novel more-powerful and low toxic antioxidants comprises several steps, including rational design strategies, theoretical methods for investigation of structure activity relationship, evaluation of the kinetic and the stability of the observed properties for prolonged periods of time and the effects of the environment condition on the radical scavenging activity [2,3]. The two most commonly used strategies adopted from the research groups focused on the design of novel antioxidants are to structurally modify existing antioxidants (ascorbic acid, α-tocopherol) in order to surpass their properties or to design an entirely new molecule [4].

The present investigation includes the evaluation of the antioxidant activity (AOA) of three new designed isonicotinoyl hydrazones – salicylaldehyde isonicotinoylhydrazone (SIH) derivatives. Hydrazones are widely studied biologically important compounds famous for a broad spectrum of biological activities - i.e. antibacterial, anticancer, anti-inflammatory, antioxidant etc [5,6]. The wide variety in the observed pharmacological properties in combination with their relatively easy and cheap methodology of synthesis has made them considerably interesting for medicinal chemists in their attempt to design concepts and synthesize novel multifunctional molecules meeting the requirements of chemical and pharmaceutical industries. The initial compounds SIH is a lipophilic tridentate iron chelator possessing noticeable antioxidant activity and modest cytotoxicity against different type of neoplastic cell lines [7]. However, the performed investigations concerning its stability in aqueous milieu have revealed rapid hydrolysis of its hydrazone bond [7,8].

For our investigation we have chosen three structurally similar isonicotinoyl hydrazones with proved cytotoxicity – two methoxy derivatives and one bromo-substituent. The previous investigations of the research group which designed the compounds studied by us - 5-bromosalicylaldehyde isonicotinoylhydrazone (5BrSIH), 3-methoxysalicylaldehyde isonicotinoylhydrazone (3mSIH) and 4-methoxysalicylaldehyde isonicotinoylhydrazone (4mSIH) includes confirmation of their structure by elemental and thermo-gravimetric analysis, IR, ^{1}H-NMR, ^{13}C-NMR spectroscopy and evaluation of the compounds with the Lipinski's rule of five [9,10].

The study has two primary interrelated aims. The first one is to determine the impact of three new hydrazone derivatives in *in vitro* systems used to investigate free radical processes. For this purpose, we have performed radical scavenging approach in systems containing stable free radicals (ABTS and DPPH) and measured the degree of oxidation of biologically important molecules via iron induced peroxidation in lipid containing model systems. The second goal is associated with the preservation of the established activity – we aimed to evaluate the potential of polysaccharide – based particles (attractive drug delivery systems possessing various advantages, i.e. nontoxicity and biodegradability) to act as protective carriers preserving the revealed antioxidant properties of the tested compounds.

2. Material and methods

2.1. UV-Vis spectra – the UV-Vis absorption spectra of the investigated compounds were taken with Shimadzu computerized spectrophotometer. The investigated compounds were diluted in 50 mM K_2HPO_4/KH_2PO_4 buffer, pH 7.4 [0.1 mmol/L], placed in quartz cuvettes and scanned over the wavelength range from 200 nm to 700 nm.

2.2. Determination of the total antioxidant capacity

To determine the total antioxidant capacity of the studied hydrazone derivatives we have chosen two of the most popular and commonly used spectrophotometric methods used for evaluation of the antioxidative role of a wide range of biological samples and new designed compounds e.g. ABTS and DPPH assays. They are rapid, simple and inexpensive methods based on the discoloration of the preformed stable free radicals – the antioxidant efficiency of the tested compounds is inversely proportional to the absorbance of the sample solution. The experimental evaluation of the studied properties has been performed using UV-Vis Shimadzu spectrophotometer and 2 ml disposable cuvettes. Fresh solutions of the radicals were prepared for each assay. Samples were analyzed in triplets.

2.2.1. ABTS method – The total antioxidant capacity (TAC) in the ABTS model system was determined by the method proposed by Re et al. [11]. The ABTS assay utilizes the free mono-cation radical of 2'2-azino-bis(3-ethylbenzothiazoline-6-sulphonic acid), generated by the oxidation of ABTS with potassium persulfate. The blue-green colored solution was dark stored for 1 day. The working solution was obtained by diluting the stock solution of the ABTS radical cation with buffer solution- K_2HPO_4/KH_2PO_4, pH 7.4 to produce a final solution with absorbance 0.7 ± 0.005 at 734.

Hydrazones samples were reacted with 2 cm^3 of ABTS$^{+\bullet}$ solution for 1 hour at room temperature. The discoloration of the pre-generated ABTS$^{+\bullet}$ radical was measured at 734 nm and the chemical response was compared with the one obtained under identical experimental conditions with the water soluble vitamin E analogue – Trolox.

2.2.2. Evaluation of the stability of the antioxidant properties – To evaluate the stability of the antioxidant properties of the studied hydrazones we have used the stable ABTS radical. We prepared stock solutions of the investigated hydrazones and hydrazones encapsulated in chitosan alginate particles. Chitosan and sodium alginate were selected as carriers due to their favorable biological properties [12,13]. The isonicotinoyl hydrazones were incorporated in the polysaccharide carriers by inotropic gelation. In particular, sodium alginate solution (3 mg/ml) was incubated with the respective hydrazone derivative for 30 min (ratio 1:10 with respect to polymer concentration), after which solution of calcium chloride (3.35 mg/ml) was slowly added to the mixture under gentle stirring (700 rpm). After that a solution of chitosan (0.75 mg/ml) was added drop wise in order to accomplish gelation between both polymers. The stock solutions were stored at 4 °C for thirty-day investigation period and every 5 days a comparative evaluation of the encapsulated and non-encapsulated hydrazones was performed.

2.2.3. DPPH method – the analytical procedure was performed according to Goupy et al [14]. For this experimental procedure a working purple colored solution of the DPPH radical in ethanol with absorbance of 1 at 517 nm was prepared. 2 cm^3 of the working solution of the radical were reacted with the studied hydrazones for 60 minutes at room temperature. After the incubation time the decrease in the absorbance reading of the probes was measured at 517 nm. Trolox was used again as referent compound. Results were presented as percentage from the control sample.

2.3. Determination of the protection effect of the studied derivatives during iron induced peroxidation in lipid containing model systems

The thiobarbituric acid assay was used to assess lipid peroxidation according to the method of Asakawa and Matsushita [15]. The peroxidation was induced in two alternative model systems one containing egg yolk lipoproteins and the other lecithin. Hen egg yolk homogenate was used as lipoproteins rich medium. According to the literature data the total solids content of egg yolk is generally around 50% and its major constituents are proteins 16% and lipids 32% [16]. All samples were prepared in PBS, pH 7.4, containing 1 mg (lecithin or egg yolk homogenate lipids) per mL and the studied hydrazones. In the control sample hydrazones were omitted. The peroxidation was induced by the addition of $FeCl_2$ with final concentration of 0.1 mmol/L. The samples were incubated at 37 °C for 30 min. Then 0.5 ml of 2.8% trichloracetic acid and 0.5 ml of thiobarbituric acid were added. The mixture was vortexed vigorously, heated in 100 °C water bath for 20 min and centrifuged at 3000 rpm for 20 min. The absorbance of the studied probes was measured at 532 nm. The degree damage of the lipid molecules was presented as percentage from the entreated control.

3. Results

The electronic spectra of the studied isonicotinoyl hydrazones in 50 mM K_2HPO_4/KH_2PO_4 at room temperature were recorded. The obtained data from the spectral study showed maxima in the middle and near UV region of their UV-Vis absorption spectra. We didn't observe absorption bands over the wavelength range where DPPH and ABTS radicals are quantified. The lack of overlapping absorbance bands makes both methods suitable for the determination of the antioxidant properties of the investigated compounds.

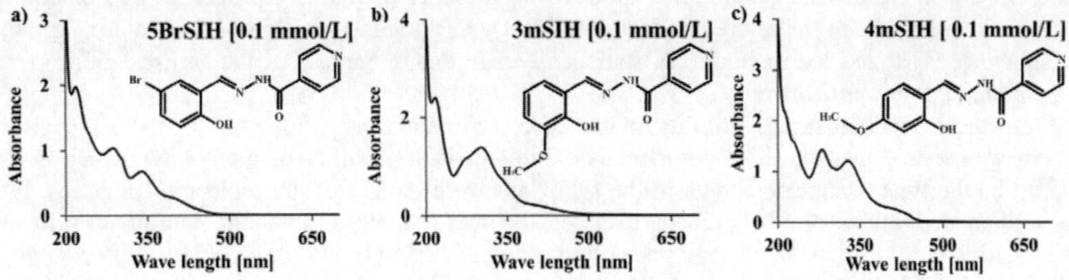

Figure 1. Absorption spectra of the studied isonicotinoyl hydrazone derivatives diluted in 50 mM K_2HPO_4/KH_2PO_4 buffer pH 7.4, over the wavelength range from 200 nm to 700 nm.

The newly designed compounds were screened for their antioxidant potential employing the *in vitro* DPPH and ABTS free radical scavenging assays. Using these methods, we have determined their capacity to quench both preformed stable free radicals utilizing Trolox as a standard. A close survey of the obtained results indicates a different magnitude of the revealed antioxidant properties against both radicals during this approach. This could be explained by the different mechanisms of reduction.

The radical scavenging activity of the SBH derivatives against ABTS has been assessed in the concentration range from 0 to 9 µmol/L (figure 2). Results are presented as AOA and the reducing capacity in terms of C-50 [µmol/L] was calculated in order to compare the antioxidant properties. C-50 is the concentration of the hydrazone that provides 50% value of AOA.

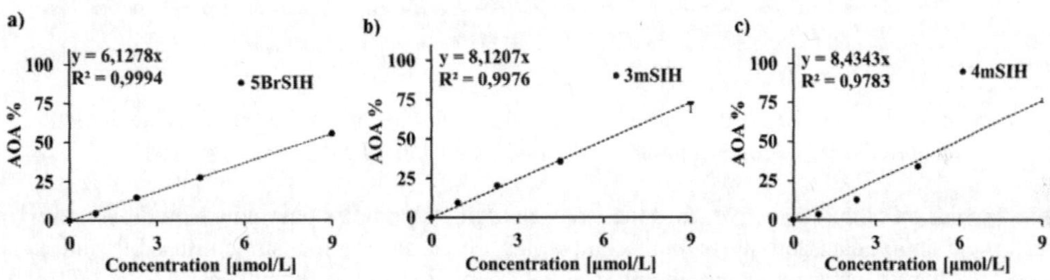

Figure 2. Antioxidant activity of the isonicotinoyl hydrazones in the ABTS model system.

All the derivatives revealed themselves as excellent antioxidants. The AOA effect rises with the increase of the concentrations of the tested hydrazone ($R^2 > 0.97$). In the whole studied concentration range the compounds possessing methoxy group as substitutes in the salicylaldehyde moiety demonstrated better radical scavenging activity correlating with bigger AOA values than the bromo-substituent.

Table 1. Results for C-50 values calculated on the basis of the experimental results obtained in the ABTS model system.

Compound	*C-50 (µmol/L*
5BrSIH	8.16 ±0.005
3mSIH	6.16 ±0.015
4mSIH	5.93 ±0.128
Trolox	8.53 ± 0.18

The comparison of the calculated C-50 values revealed better antioxidant properties of isonicotinoyl hydrazones compared to the reference Trolox (table 1). Varying the position of the methoxy group in the salicylaldehyde has led to slight but statistically significant increase of the antioxidant properties corresponding to lower C-50 value.

From the evaluation of the stability of the disclosed antioxidant properties in the ABTS model system we can see that the three compounds exhibit variable stability of their AOA which doesn't depend on the type of the substitutes in the salicylaldehyde moiety of the molecule (figure 3). From the studied derivatives 3mSIH present itself as the more stable antioxidant. During the 30 days investigation period the AOA of its stock solution diminished only by 13.08%. The AOA activity of the other methoxy bearing isonicotinoyl hydrazones (4mSIH) and the bromo - substituent (5BrSIH) diminished respectively by 75.12% and 56.44%. During the whole investigated period 3mSIH stock solution is the most potent antioxidant from the studied compounds.

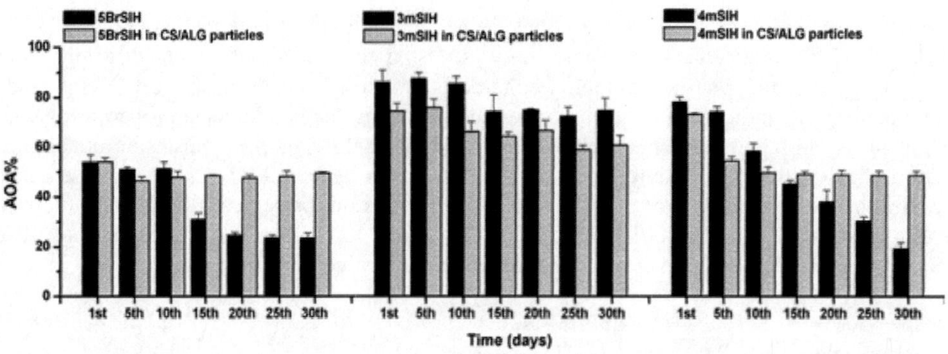

Figure 3. Changes of the AOA of non-encapsulated and encapsulated into polysaccharide particles hydrazones against ABTS radical during 30 days storage period.

The incorporation of the derivatives into the chitosan-alginate particles has had a beneficial preserving effect on the antioxidant activity of both compounds which have demonstrated unstable antioxidant properties. This effect is more prominent in the case of the bromo-substituent. At the end of the storage period the encapsulated 5BrSIH has an antioxidant activity approximately two times higher compared to the initial solution of the non-encapsulated hydrazone. During the studied period the reduction of the AOA was only by 9.17%.

From the studied derivatives 4mSIH stock solution was the less stable. The reduction in its AOA for the investigated period was 75.13%. We also observed significant diminishment of the AOA of the loaded in polysaccharide particles 4mSIH – 33.51%. At the end of the investigated period the encapsulated 4mSIH exhibit two and a half times higher activity compared to the non-encapsulated compound solution.

The comparison of the reduction of the activity of the encapsulated and non-encapsulated 3mSIH has disclosed an almost equal stability of both solutions and diminishment of the AOA by less than 20%.

The obtained results from the model system containing the stable DPPH radical are shown in figure 4. Due to the poor reactivity of the studied hydrazones in this model system only data concerning the highest investigated concentration [90 μmol/L] are presented.

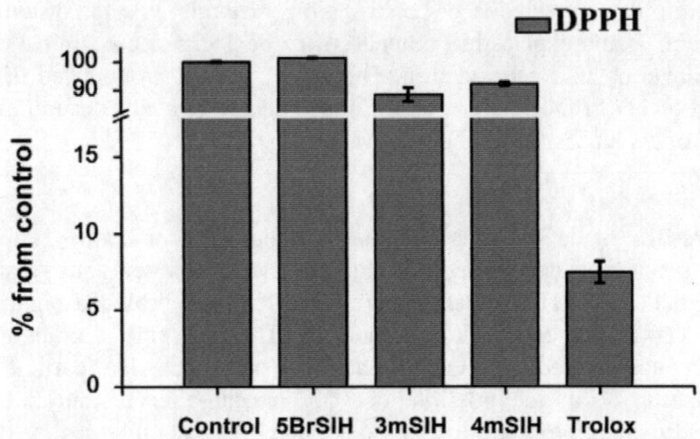

Figure 4. DPPH radical scavenging activity of the studied hydrazones and the reference Trolox.

The sample containing the bromo - substituent (5BrSIH) has the same absorbance as the control which indicates impossibility for hydrogen atom abstraction from its molecule and formation of hydrazonyl radical. The two methoxy bearing isonicotinoyl hydrazones displayed a slight but statistically significant increase of the investigated in the DPPH system properties. This effect is better expressed in the sample containing 3mSIH which has demonstrated the highest DPPH scavenging activity among the synthesized compounds.

Even in this case the degree of decolonization and the exhibited DPPH radical scavenging properties are very weak compared with the one of the reference Trolox. At a concentration of 46 μmol/L Trolox induced almost complete quenching efficacy and full decolonization of the DPPH solution.

In the model system of iron induced lipid peroxidation the isonicotinoyl hydrazones demonstrated different extent of protection of the molecules depending on the used substrate (figure 5).

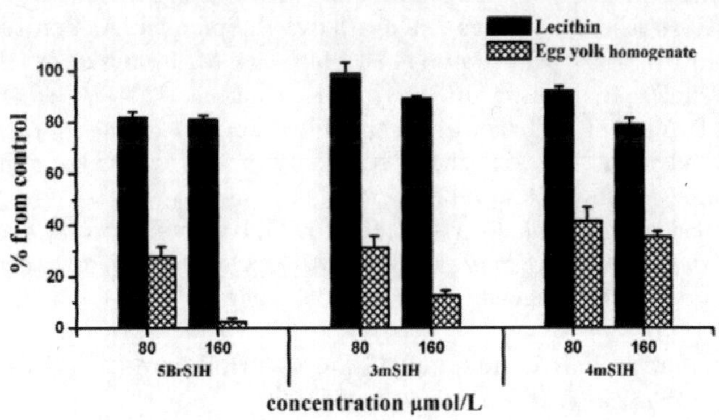

Figure 5. Degree of protection of the studied hydrazones during iron induced peroxidation in lipid containing model systems.

In the lecithin containing system the bromo-substituent (5BrSIH) had an identical effect on both tested concentrations. The methoxy derivatives exhibited a similar but statistically different concentration dependent activity. The variation of the position of the methoxy group from 3rd to 4-th position had a slight beneficial effect on the evaluated in the system effect.

The data from the iron induced peroxidation methods in the system containing egg yolk homogenate which is more plausible compared to the lecithin containing one uncover significant decrease of the level of lipid peroxidation. This effect is most pronounced in the samples containing the bromo-substituent (5BrSIH) were at the highest investigated concentration of 160 μmol/L we established lack of evidences for lipid oxidative damage.

5. Conclusion

The obtained results made evident that the investigated isonicotinoyl hydrazones have proven themselves as promising candidates for further evaluation as antioxidant agents. They have demonstrated excellent ABTS radical scavenging effect and provided significant protection in the model systems containing egg yolk homogenate. The performed comparative evaluation of the stability of non-encapsulated and encapsulated in polysaccharide particles hydrazones revealed valuable information about the possible use of chitosan-alginate particles as protective carriers preserving the antioxidant properties of the compounds. The stability of the AOA of the encapsulated bromo-substituent (5BrSIH) along with the data from the iron induced lipid peroxidation denote that despite its lower reactivity against stable free radicals the structural modification associated with incorporation of bromine atom in the salicylaldehyde moiety of the isonicotinoyl hydrazones deserve more detailed analysis.

Acknowledgements
Financial support by the Medical University of Sofia (Council of Medical Science, Contract no. 38/2016) is gratefully acknowledged.

References

[1] Florence TM 1995 *Aust N Z J Ophthalmol.* **23** 3
[2] Thomas CE 1997 *Handbook of synthetic antioxidants.* New York: Marcel Dekker; 1
[3] Espinoza-Fonseca L.M 2006 *Bioorg. Med. Chem.* **14** 896
[4] Zhang HY 2005 *Curr. Comput.-Aided Drug Des* **1** 257
[5] Rollas S, Küçükgüzel SG 2007 *Molecules* 12 1910
[6] Narang R, Narasimhan B, Sharma S 2012 *Curr Med Chem.* **19** 569
[7] Potůčková E, Hrušková K, Bureš J, Kovaříková P, Špirková IA, Pravdíková K, Kolbabová L, Hergeselová T, Hašková P, Jansová H, Macháček M, Jirkovská A, Richardson V, Lane DJ, Kalinowski DS, Richardson DR, Vávrová K, Šimůnek T 2014 *PLoS One.* **13** 9
[8] Richardson D, Vitolo LW, Baker E, Webb J 1989 *Biol Met* **2** 69
[9] Nikolova-Mladenova B, Halachev N, Iankova R, Momekov G, Ivanov D, 2011 *Arzneimittelforschung/Drug research,* **61** 714
[10] Nikolova-Mladenova B, Bakalova A, Momekov G, Ivanov D *International Scientific Journal of Medical and Biological Sciences* http://bioscience.scientific-journal.com
[11] Re R, Pellegrini N, Proteggente A, Pannala A, Yang M, Rice-Evans C 1999 *Free Radic Biol Med.* **26** 1231
[12] Kumar MN, Muzzarelli RA, Muzzarelli C, Sashiwa H, Domb AJ 2004 *Chem Rev.* **104** 6017-84
[13] Baldrick P 2010 *Regul Toxicol Pharmacol.* **56** 290
[14] Goupy P, Dufour C, Loonis M, Dangles O 2003 *J Agric Food Chem.* **51** 615
[15] Asakawa T, Matsushita. S 1980 *Lipids* **1** 137
[16] Ahn D, Kim SM, Shu H 1997 *Poult Sci.* **76** 14

Depot effect of bioactive components in experimental membrane filtrations

D Mitev[1], D Peshev[2], G Peev[2] and L Peeva[3]

[1] REGPOT-2012-2013-1 NMP INERA Project, Institute of Solid State Physics-BAS, 72 Tzarigradsko Chaussee Blvd., 1784 Sofia, BULGARIA
[2] University of Chemical Technology and Metallurgy, Department of Chemical Engineering, 8 Kl. Ohridsky Blvd., 1756 Sofia, Bulgaria
[3] Chemical Engineering and Chemical Technology, Imperial College London, Exhibition Road, London SW7 2AZ, U.K.

Email: dpmitev@yahoo.com

Abstract. Depot effects were found to be accompanying phenomena of membrane separation processes. Accumulation of target species in the membrane matrix during feasibility tests can hamper proper conclusions or compromise the filtration results. Therefore, we investigated the effects of delayed membrane release of chlorogenic acid and caffeine, considered as key compounds of interest in spent coffee products' recovery treatment. Permeate fluxes and key components release were studied in course of 24 hours via nanofiltration of pure solvent, both immediately after the mock solution filtration and after idle stay. Conclusions are drawn and recommendations advised for proper analysis of experimental data on membrane screening.

1. Introduction

Adsorptive depot effects have drawn significant attention in the field of membrane filtration, especially when small scale experiments were involved. Ideally, no part of the analytes or the compounds of interest should be retained in the membrane matrix. However, in real applications, considerable amount of the filtered solutes can be adsorbed during the initial filtration stages, predetermined by the membrane (material, presence/absence of wetting agents, pore size and distribution), solute type, pH, ionic strength, and even filtration flow rate [1]. A range of separations have been studied in the scope of this phenomenon: from individual compounds such as toluidine blue [1], benzalkonium chloride, chlorhexidine acetate, phenylmercuric nitrate and phenylethyl alcohol [2-3], estrogen 17β-estradiol [4], triethanolamine and sodium benzoate [5], a gamut of sixteen neutral, acidic and basic compounds [6], surfactants [7], to viruses [8] and bovine serum albumin (BSA) [9] and other peptides (luteinizing hormone releasing hormone, gamma globulin and BSA) [10], monoclonal antibodies and surfactant influence thereon [11]; with membranes of different materials being mainly of microfiltration types (sterilising grade) though.

This research reveals the depot effects accompanying nanofiltration of standard caffeine (CAF) and chlorogenic acid (CGA) solutions, aimed at modeling of spent coffee ground recovery via membrane filtration.

2. Experimental part

2.1. *Materials*

All reagents used were of analytical grade quality: chlorogenic acid (Sigma-Aldrich Chemie GmbH, Steinheim, Germany), caffeine (Chimspectar, Sofia, Bulgaria). Distilled water (water still GFL Typ 2004, Burgwedel, Germany) was used throughout the work.

2.2. *Instruments*

Following instrumentation was used throughout the work:

Membrane filtration. "Dead-end" filtration cell of type "METcell" (Membrane Extraction Technology Ltd., London, UK), fitted with 0.0054 m^2 nanofiltration membrane "Microdyn Nadir" NP 030 P (Microdyn-Nadir GmbH, Wiesbaden, Germany); operating pressure up to 60 bar (compressed nitrogen) under internal stirring 1000 min^{-1}.

Spectrophotometry. An S-22 UV/Vis type of spectrophotometer (Boeco, Germany) was used for determination of the maximal spectral absorption at wavelengths of 275 and 330 nm.

2.3. *Preparation of the standard solutions*

Feed solutions of 497.01 mg kg^{-1} caffeine and 495.70 mg kg^{-1} chlorogenic acid were prepared with distilled water. Given concentrations correspond to the spectrophotometric evaluation of the final solutions, according to the standard curves represented on figure 1 (a-b).

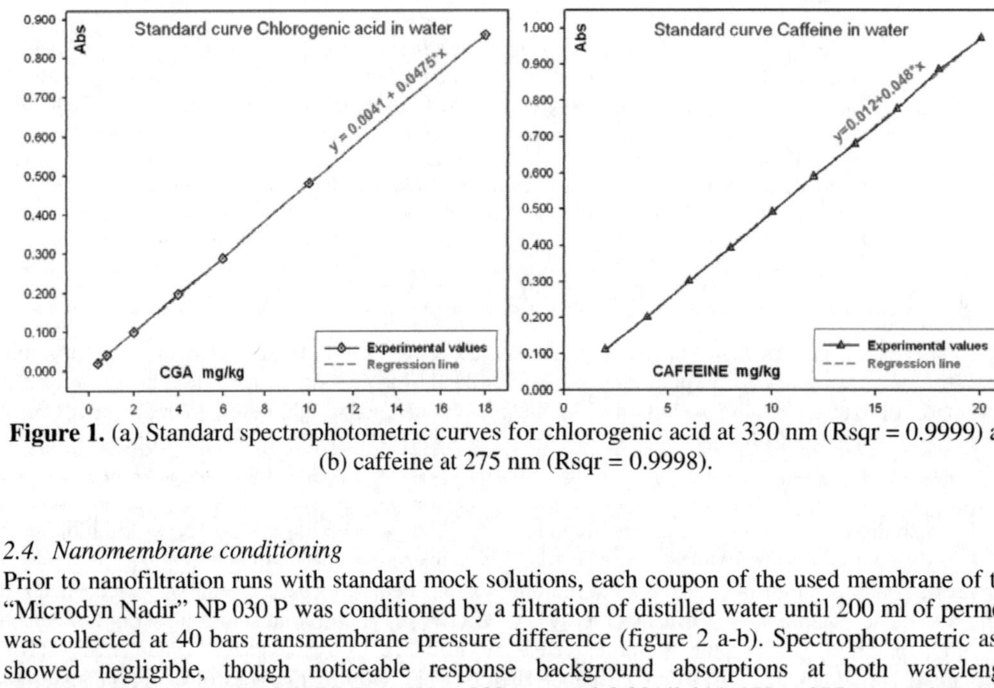

Figure 1. (a) Standard spectrophotometric curves for chlorogenic acid at 330 nm (Rsqr = 0.9999) and (b) caffeine at 275 nm (Rsqr = 0.9998).

2.4. *Nanomembrane conditioning*

Prior to nanofiltration runs with standard mock solutions, each coupon of the used membrane of type "Microdyn Nadir" NP 030 P was conditioned by a filtration of distilled water until 200 ml of permeate was collected at 40 bars transmembrane pressure difference (figure 2 a-b). Spectrophotometric assay showed negligible, though noticeable response background absorptions at both wavelengths (retentate/total permeate): 0.015/0.011 AU at 330 nm; and 0.021/0.011 AU at 275 nm, which would correspond to ±0.229/±0.251 mg kg^{-1} CGA; and ±0.188/±0.375 mg kg^{-1} CAF respectively if they were really present.

Corresponding response of the NP 030 P nanomembrane shows up at different pressures during the water conditioning step. Conditioned at lower pressure membranes express higher permeate fluxes, which are fully reversible after moderate pressure changes, as shown by the 5 to 10-bars plot on figure 2(b). A preliminary conditioning at 40 bars immediately before 10 bars water filtration lowers the flux's plateau to ~4.5 LMH/bar though; this in turn might be an indicator of structure transformations of the membrane after the high pressure treatment. Despite these differences it must be noted that all permeate fluxes obeyed the manufacturers specification of >1 LMH/bar.

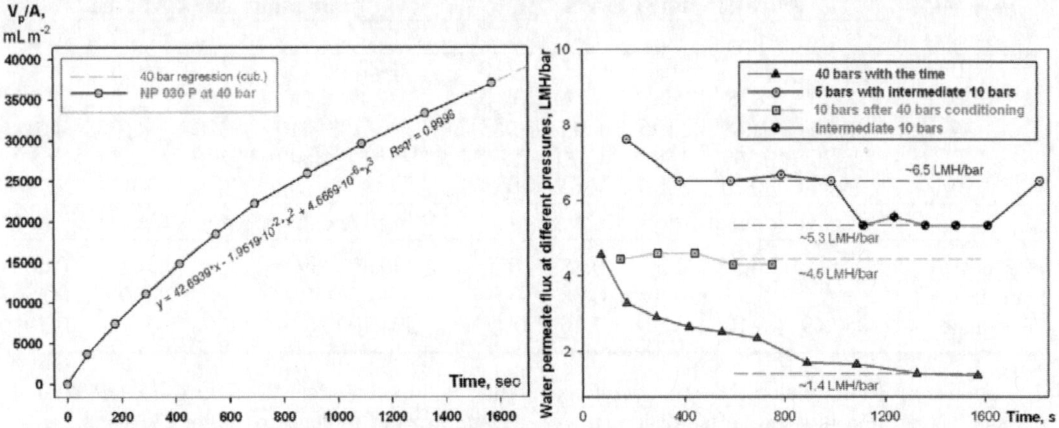

Figure 2. Kinetics of the NP 030 P nanofiltration conditioning: (a) cumulative permeate volume per unit membrane area versus time at 40 bars; (b) water permeate fluxes of NP 030 P at different pressures versus time.

2.5. Nanomembrane filtration

100 ml of the standard solutions were subjected to nanofiltration at temperature of 298.15 K and pressure of 30 bar. The membrane separation continued until 51 ml of permeate (P) were isolated for CGA and 52 ml for CAF; with retentate (R) being the rest.

3. Results and Discussion

The aim of this research is to establish the separation performance in terms of permeate flux and solute rejection of the nanomembrane in water solutions of CGA and CAF, while accounting for and quantifying the related depot-effects in the NP 030 P membrane.

3.1. Nanomembrane filtration of standard solutions

Nanofiltration runs of both standard solutions were performed in "dead-end" filtration mode, under intensive mixing provided by internal magnetic stirrer as described in Section 2. Furthermore, four subsequent filtrations of 100 ml distilled water each one were done in both cases. At every filtration 50 ml of permeate were collected, followed by additional 3 ml permeate for estimation of C_{pe} (permeate concentration at the end of nanofiltration; mg kg^{-1}). The output kinetics of both the filtration procedures and "depot" adsorption values are shown in table 1 and figure 3 (a-b).

"Microdyn Nadir NP 030 P" is a commercial nanofiltration membrane (30 % retention for NaCl and 80-95 % for Na_2SO_4; MWCO ~400 Da); with polyethersulfone (PES) active membrane layer on PE/PP support [12]. During the filtrations considerable rejection was registered for CGA (molecular weight 354.31 Da), where a quadratic regression was the best fit for the nanofiltration kinetics (figure 3(a)). The membrane rejection in regard to the CGA contents of the standard solution was calculated according to [13]:

$$R_{CGA} = (lg(C_r / C_f) / lg(V_f / V_r))\ .\ 100 = 75.7\ \% \qquad (1)$$

Table 1. Kinetics of separating filtration (T), membrane washings (t_i) and concentration of the respective fractions.

Permeate,	**Filtration time CGA**, sec.					**Filtration time CAF**, sec.				
ml	T_{Fcga}	t_{1cga}	t_{2cga}	t_{3cga}	t_{4cga}	T_{Fcaf}	t_{1caf}	t_{2caf}	t_{3caf}	t_{4caf}
10	540	450	424	333	330	146	152	130	135	134
20	1130	835	805	669	640	273	284	260	280	300
30	1780	1248	1196	1017	965	404	410	390	410	470
40	2430	1668	1580	1390	1275	550	547	540	555	665
50	3145	2110	1966	1765	1600	704	693	689	705	863
	Active concentrations CGA, mg kg^{-1}					**Active concentrations CAF**, mg kg^{-1}				
Permeate C_p	58.95	25.35	2.39	0.88	0.31	440.48	59.45	25.60	0.21	0.60
Permeate C_{pe}	-	3.93	0.84	0.38	0.04	-	1.98	0.08	0.06	0.35
Retentate	850.49	6.10	0.61	1.26	0 (neg.)	504.19	0.67	0.08	0 (neg.)	0.50

The obtained rejection value shows moderate capability of the membrane to retain CGA and related compounds, present in spent coffee residues. As far as in terms of molecular weight CGA represents the lower limit amongst the antioxidant constituents in coffee raw material [14], one shall expect even higher efficiency of the investigated membrane in concentration of antioxidant liquid extracts from coffee.

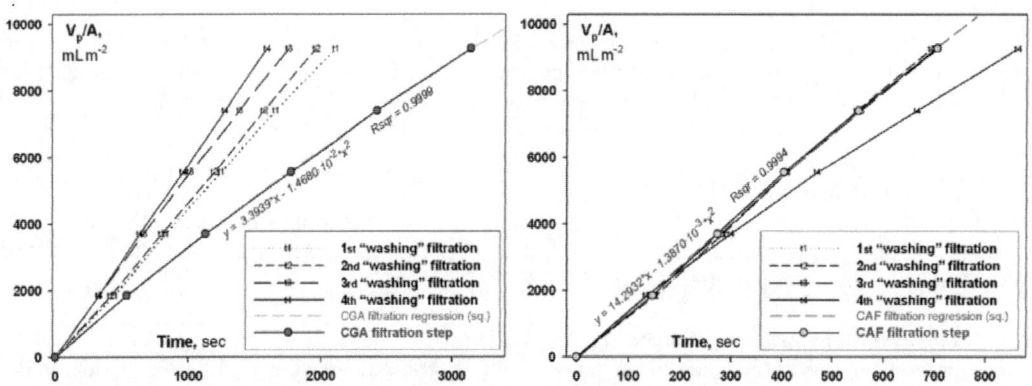

Figure 3 (a) - (b). NP 030 P nanofiltration of CGA and CAF solutions, plus subsequent four pure water (washing) filtrations in each case.

The experimental results showed negligible retention capacity of the membrane with respect to caffeine (molecular weight 194.19 Da; figure 3 (b)):

$$R_{2\,CAF} = (lg(C_r / C_f) / lg(V_f / V_r)) \,.\, 100 = 2.0\ \% \qquad (2)$$

The combination of high rejection for CGA and higher molecular weight polyphenols with antioxidant activity and negligible rejection for CAF at the same time, demonstrates the membrane applicability in production of caffeine free coffee products as well as for isolation of natural caffeine. According to the material balance 4.9 mg or 9.9 % of the available CGA was adsorbed onto the membrane during the nanofiltration step; respectively 2.6 mg or 5.2 % for the nanofiltration of CAF (table 1). These values indicate the uncertainty in the determination of the membrane rejection values, using laboratory scale batch experiments, especially in case of new "unsaturated" membranes.

3.2. The effect of "washing" filtrations

In the case of CGA, desorption was a relatively slow process and the influence of concentration polarisation was noticeable until the third washing filtration (quadratic regression fit). With the course of washings the process fastened evenly, despite that between the second and third "washing" filtrations an 18-hour-long pause took place (figure 3(a)).

For the CAF nanofiltrations, relatively small molecules did not block the membrane; washing filtrations practically overlap the nanofiltration kinetics. The only deviation here is the course of the last "washing" filtration, made after an 18-hours-long pause as well. Possible reasons for the phenomenon include a relaxation change of the membrane, or a release of dissolved CAF by the membrane polymer after the idle period. Both hypotheses need a further confirmation though

4. Conclusions

Depot effect of CGA and CAF water solutions nanofiltration have been established (those of CGA being significantly stronger), which has to be considered evaluating process characteristics.

A nanofiltration step was suggested, with potential integration in technologies for production of decaffeinated alkaloid beverages, or as a separation method for chlorogenic acid or caffeine from suitable sources.

Acknowledgements

REGPOT-2012-2013-1 NMP INERA Project is kindly acknowledged. The authors also acknowledge University of Chemical Technology and Metallurgy – Sofia for financial support of this research (Project No. 11572).

References

[1] Cooney DO 1992 Adsorption of organic solutes on membrane filters during aqueous phase filtration. I. Basic rate and equilibrium studies using toluidine blue *Separation Science and Technology* **27** 2001-2019

[2] Bin T, Kulshreshtha AK, al-Shakhshir R and Hem SL 1999 Article I. Adsorption of benzalkonium chloride by filter membranes: mechanisms and effect of formulation and processing parameters *Pharmaceutical Development and Technology* **4**(2) 151-165

[3] Naido NT, Price CH and McCarthy TJ 1972 Preservative Loss from Ophthalmic Solutions During Filtration Sterilization *Australian Journal of Pharmaceutical Sciences*, NS **1**(1) 16-18

[4] Min L, Lai, EPC and Yu Y 2012 Removal of 17 β-Estrodiol by nylon filter membrane: Adsorption kinetics and thermodynamics *International Journal of Research & Reviews in Applied Sciences*, **11**(1) 67-73

[5] Kohyama T 2012 The filter adsorption mechanism in photoresist materials *SPIE Advanced Lithography* **0001** 83252I-83252I-11 doi:10.1117/12.916082

[6] Carlson M and Thompson RD 2000 Analyte Loss Due to Membrane Filter Adsorption as Determined by High-Performance Liquid Chromatography *Journal of Chromatographic Science* **38**(2) 77-83

[7] Zhou JX, Qiu J, Jiang G, Zhou C, Bingham N, Yeung H, Dransart B, Wadhwa M and Tressel T 2008 Non-specific binding and saturation of Polysorbate-20 with aseptic filter membranes for drug substance and drug product during mAb production *Journal of Membrane Science* **325**(2) 735-741

[8] Cliver DO 1968 Virus interactions with membrane filters *Biotechnology and Bioengineering* **10**(6) 877–889

[9] Roesink HDW, Beerlage MAM, Potman W, van den Boomgaard Th, Mulder MHV and Smolders CA 1991 Characterization of new membrane materials by means of fouling experiments Adsorption of bsa on polyetherimide-polyvinylpyrrolidone membranes *Colloids and Surfaces* **55** 231-243

[10] Van den Oetelaar PJM, Mentink IM, Brinks GJ 1989 Loss of Peptides and Proteins Upon Sterile Filtration Due to Adsorption to Membrane Filters Drug Development and Industrial Pharmacy 15(1) 97-106

[11] Mahler HC, Huber F, Kishore RS, Reindl J, Rückert P and Müller R 2010 Adsorption behavior of a surfactant and a monoclonal antibody to sterilizing-grade filters *Journal of Pharmaceutical Sciences* **99**(6) 2620-2627

[12] NADIR® NP030 P - Membrane for Nanofiltration - Data Sheet 2016 (September) http://www.lenntech.com/Data-sheets/Microdyn-Product-catalogue-L.pdf

[13] Peev G, Penchev P, Peshev D, Angelov G 2011 Solvent extraction of rosmarinic acid from lemon balm and concentration of extracts by nanofiltration: Effect of plant pre-treatment by supercritical carbon dioxide *Chemical Engineering Research & Design* **89** 2236-2243

[14] All polyphenols found in Coffee beverage – Phenol-explorer. Database on polyphenol content in foods – 2016 (September); http://phenol-explorer.eu/contents/food/552

Synergetic effect of TiO_2 nano filler additives on conductivity and dielectric properties of PEO/PVP nanocomposite electrolytes for electrochemical cell applications

H K Koduru[1,3*], K K Kondamareddy[2], M T Iliev[3], Y G Marinov[3], G B Hadjichristov[3], D Karashanova[4] and N Scaramuzza[1,5]

[1] Dipartimento di Fisica, Università degli Studi della Calabria, Via P. Bucci, Cubo 33B - 87036 Rende (CS) – Italy.
[2] School of Physics and Technology, Key Laboratory of Artificial Micro/Nano Structures of Ministry of Education, Wuhan University, Wuhan, 430072, China.
[3] Georgi Nadjakov Institute of Solid State Physics, Bulgarian Academy of Sciences, 72 Tzarigradsko Chaussee, Sofia 1784, Bulgaria.
[4] Institute of Optical Materials and Technologies Bulgarian Academy of Sciences Acad. Georgy Bonchev St., bl. 109 1113 Sofia, Bulgaria.
[5] CNISM – Unità di Ricerca di Cosenza c/o Dipartimento di Fisica, Università degli Studi della Calabria Via P. Bucci, Cubo 31C - 87036 Rende (CS) – Italy.

* E-mail address: koduruharika@gmail.com

Abstract. Sodium-ion conducting PEO/PVP blend based solid polymer electrolyte films complexed with $NaIO_4$ salt and nano-sized TiO_2 fillers are fabricated by employing a solution casting technique for Na-ion battery applications. Measurements of X-ray diffraction (XRD) and thermogravimetric analysis (TGA) are carried out to investigate the crystallinity and thermal stability of the solid polymer electrolytes. Scanning electron microscopy (SEM) and Transmission electron microscopy (TEM) studies are performed to understand the modifications in surface morphological features and to evaluate the size and distribution of dispersed nano-sized TiO_2 fillers. The room temperature ionic conductivities of polymer electrolyte films are investigated by impedance analysis in the frequency range 1 MHz - 1 Hz. The nano-sized TiO_2 (3 wt%) filled composite electrolyte of 'PEO/PVP/$NaIO_4$ (10 wt%)' demonstrates a maximum room temperature conductivity of 9.82 X 10^{-6} S/cm. The influence of TiO_2 filler on conductivity and dielectric properties are presented in this report.

1. Introduction

Although lithium ion batteries (LIBs) have been recognized as a promising energy storage devices [1-3] their high cost, less abundance, environmental impact and safety limitations impede the widespread implementation of lithium insertion materials in future battery technologies [4-8]. Therefore, there is urgent necessity to search for alternative energy storage system technology capable of complementing the current Li-ion battery technology. Amongst the accessible battery chemistries, sodium (Na) based rechargeable batteries (SIBs) have recently captured much attention because they are environmentally friendly, non-toxic, low cost and abundant materials [9, 10]. Substantial efforts of research, development, and demonstration are currently in progress to replicate the performance of well-established Li-ion batteries using sodium ion batteries [11-13]. Solid polymer electrolytes (SPEs) have attracted intensive studies during the past decades due to their potential applications in various advanced devices including metal ion batteries, supercapacitors, and stretchable actuators. Compared to the traditional organic liquid electrolytes, SPEs offer enhanced safety, stability and thin film manufacturability but their low ionic conductivity especially at room temperature has suppressed their development [14-16].

Polyethylene oxide (PEO) based polymer electrolytes have been employed extensively for battery applications since Armand et al. have demonstrated the feasibility of using PEO as a potential ion conductive electrolyte [17]. PEO is one of the widely investigated host polymers used for synthesizing SPEs to employ in alkali metal ion-conducting batteries due to its high electrochemical stability, good solvation, complexation and ion dissociation abilities [18-25]. Several researchers have reported about various sodium salt complexes of PEO based polymer electrolytes for sodium ion battery applications [26, 27]. However based on previous reports, the semicrystalline nature of PEO at room temperature, subsequently limits its ionic conductivity [28]. The most straight forward approach to overcome this problem is modifying the PEO matrix in order to decrease its degree of crystallinity. One of the most promising alternate choices of enhancing the amorphous phase in PEO based electrolyte systems is blending of PEO with suitable higher amorphous polymer [29-32]. Polyvinyl pyrrolidone (PVP) has been identified as a compatible partner to PEO which exhibits higher order of amorphosity. PVP has high glass transition temperature, mechanical and thermal stabilities, provided its carbonyl group (C = O) enables to form different complexes of alkali metal ion salts at wide range of concentrations. In particular, reinforcement of nano-sized materials with polymer electrolyte systems can modify their microstructural properties and enhance their conductivity properties [33]. Several researchers have reported about micro-or nano-sized ceramic fillers doped sodium-ion polymer electrolytes, such as ZrO_2, $BaTiO_3$,Sb_2O_3,Al_2O_3 or silica nanoparticles [34, 35]. In the present report, $NaIO_4$ salt complexed and nano-sized TiO_2 fillers doped PEO/PVP blend based electrolyte systems are prepared by the solution cast technique and investigated the influence of nano-sized TiO_2 fillers on conductivity and dielectric properties of salt complexed PEO/PVP blends solid polymer electrolytes.

2. Experimental

PEO and PVP of molecular weights of 5 X 10^6 and 3.6 X 10^5 respectively, were procured from Aldrich and employed without any further purification to prepare PEO/PVP solid state blend electrolytes. Sodium periodate salt ($NaIO_4$, Sigma Aldrich) and Titanium dioxide nano powder (TiO_2, Sigma Aldrich) were used as additives and methanol (Aldrich) was used as a solvent. Appropriate amounts of PEO and PVP polymers were dissolved in methanol solution and followed by mechanical stirring at room temperature for 15 hours to obtain uniform mixture of blend. In the meanwhile, 10 wt% of $NaIO_4$ salt was dissolved separately in methanol solution and added to the prepared viscous PEO/PVP polymer solutions to obtain 'PEO/PVP/$NaIO_4$' polymer complex. Nano-sized TiO_2 powder (3 wt%) was dispersed in methanol solution separately, which was sonicated for 30 minutes, and added to 'PEO/PVP/$NaIO_4$(10 wt%)' electrolyte solution to prepare 'PEO/PVP/$NaIO_4$(10 wt%)/TiO_2(3 wt%)' nano composite electrolyte. The as-prepared viscous solutions were poured into polypropylene dishes and the solvent (methanol) was allowed to evaporate slowly at room temperature to harvest free-standing polymer electrolyte films. All electrolyte films were, vacuum dried at 45°C to remove traces of methanol solvent and kept in desiccators filled with silica gel desiccants for several hours before being characterized to avoid any traces of moisture.

The structural properties of the electrolytes were analysed by recording X-ray diffraction patterns (XRD) in the 2θ range of $10^o – 80^o$ with a constant step 0.02^0, counting time 35s/step on a Bruker D8 Advance diffractometer with Cu Kα radiation and Lynx Eye detector. SEM (Philips 515 digitalized) images were monitored to study the effect of complexation with $NaIO_4$ salt and additive nano-sized TiO_2 filler on surface morphological features of blend electrolytes. TEM (HR STEM JEOL JEM 2100

IOP Conf. Series: Journal of Physics: Conf. Series 780(2017) 012006 doi:10.1088/1742-6596/780/1/01200

acceleration voltage 80 - 200 kV, maximum resolution – 0.23 nm) measurements were carried out to estimate size and distribution of doped nano-sized TiO_2 fillers in the matrices of polymer electrolytes. Thermal stabilities of solid polymer electrolytes were studied by TGA carried out under nitrogen gas atmosphere by using TGA Q50 TA operating in the temperature range 303 K to 1073 K and at a heating rate of 10 ^{0}C/min. Solid polymer electrolyte films were sandwiched between two copper electrodes and ionic conductivity studies were carried out at room temperature by a.c. impedance measurements using Biologic potentiostat/galvanostat (SP – 200) in the frequency range of 1 MHz - 1 Hz.

3. Results and Discussion

3.1. Microstructural properties

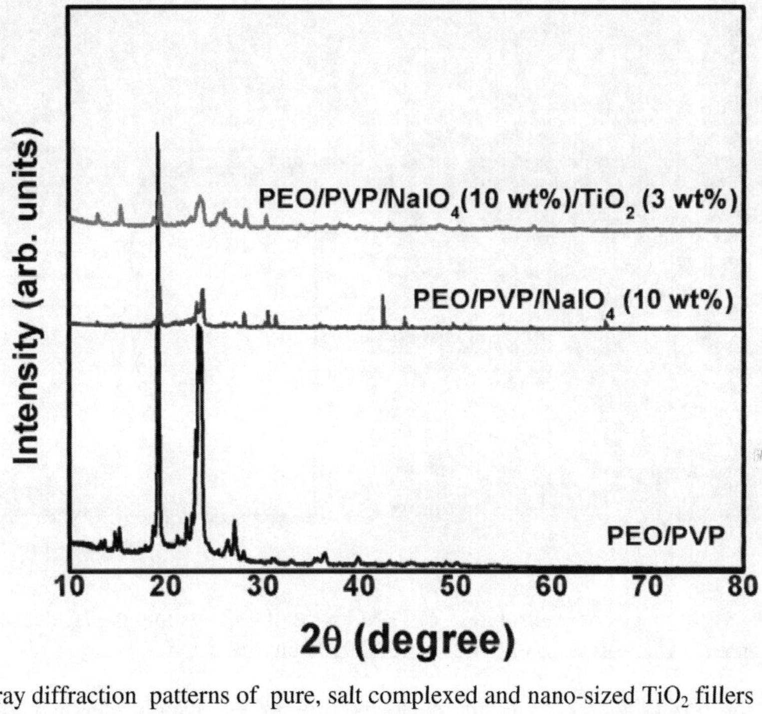

Figure 1. X-ray diffraction patterns of pure, salt complexed and nano-sized TiO_2 fillers doped electrolytes.

As evidenced from XRD measurements, PEO/PVP blend electrolyte displays intense diffraction peaks at 2θ = 19.30 and 2θ = 23.40 corresponding to characteristic crystalline peaks of PEO (as shown in figure 1) which originates from the ordering of polyether side chains and strong intermolecular interaction between PEO chains through the hydrogen bonding [36]. The intensity of characteristic diffraction peaks of PEO is noticed to be decreased as a result of addition of NaIO4 (10 wt%) salt and 3 wt% of nano-sized TiO2 filler. It evidences for further increase of amorphous portion in the matrix of the electrolytes. This favours for enhancement of mobility of the charge carriers in the polymer electrolytes and leads to increase of ionic conductivity [37]. As evidenced from SEM studies, the surface of pure PEO/PVP blend electrolyte (figure 2(a)) is observed to be relatively rough and composed with micro-void and cracks, when compared to the polymer blend composite, which signifies the crystalline nature. The formation of micro-voids over the surface of films is due to rapid evaporation of the solvent during the preparation of polymer blends. Apparently, surface smoothness of nano-composite electrolyte films is further increased as a result of incorporating nano-sized TiO2 filler in the matrix of salt complexed polymer electrolyte (figure 2(b)).

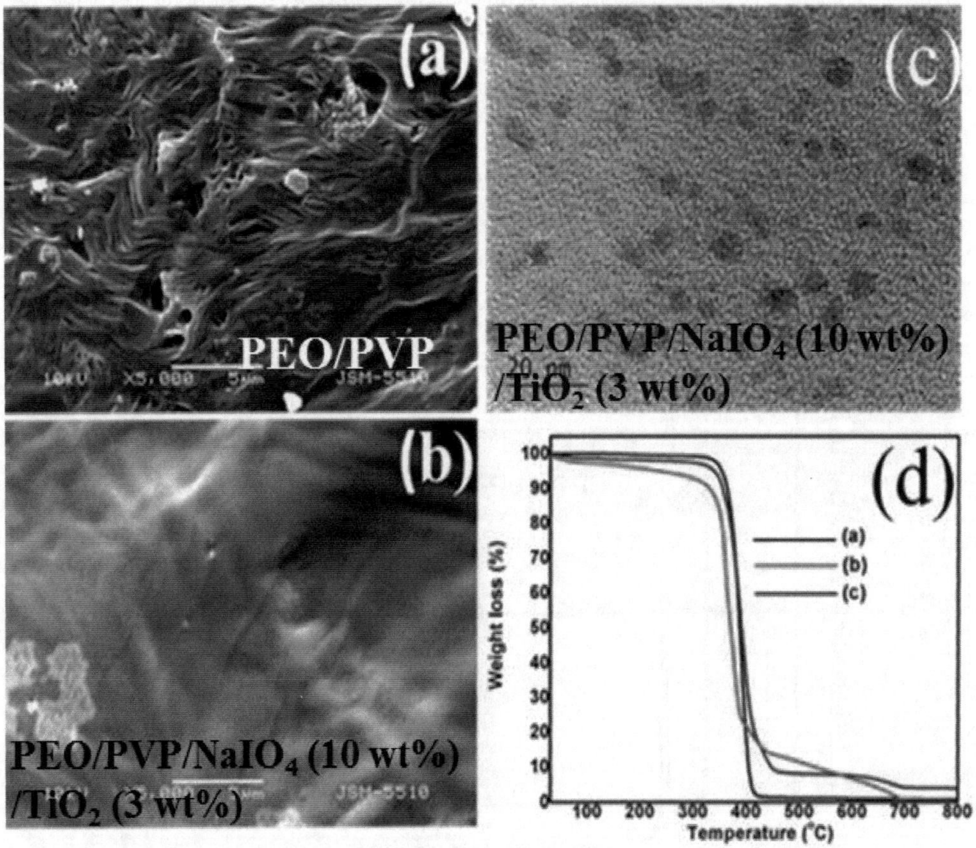

Figure 2. (a, b) SEM images of electrolytes, (c) TEM image of TiO_2 nano particles doped salt complexed electrolyte, (d) TGA profiles of different electrolytes.

It evidences for further increase of amorphous portion in the matrix of salt complexed nano composites, which favours for the enhancement of ionic conductivity [38-40]. From the TEM image (figure 2(c)) the average size of nano fillers is found to be 10 nm and dispersed nano particles are observed to be uniformly distributed in the matrix of salt complexed blend electrolytes. The nano-composite polymer electrolytes should have good thermal stability at high temperatures to meet the criteria for employing them as solid electrolytes in sodium ion batteries. Figure 2d represents TGA curves for pure PEO/PVP blend (black line), $NaIO_4$(10 wt%) salt complexed PEO/PVP blend (red line) and nano composite 'PEO/PVP/$NaIO_4$(10 wt%)' (blue line) electrolytes. The first weight loss region at lower temperatures (< 100 ^{0}C) originates from the loss of water absorbed during the loading of samples and a major weight loss region at higher temperatures (> at above 300 ^{0}C) is associated with the decomposition of pure PEO and nano composite matrices [41]. Evidently, nano-sized TiO_2 filler (3 wt%) doped PEO/PVP/$NaIO_4$(10 wt%) polymer electrolyte is demonstrated appreciable thermal stability.

3.2. Conductivity and Dielectric properties

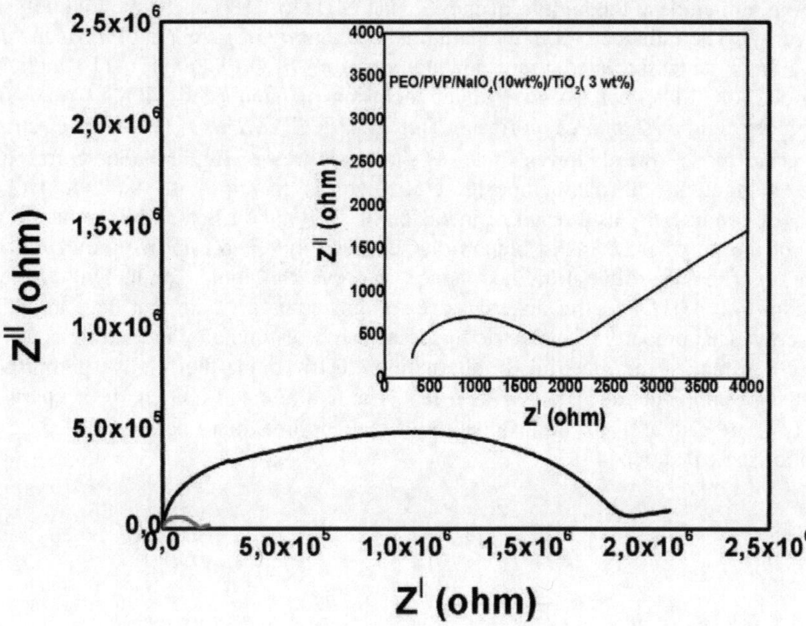

Figure 3. Impedance plots for PEO/PVP (black line), 'PEO/PVP/NaIO_4(10 wt%)' (red line) and 'PEO/PVP/NaIO4(10 wt%)/TiO_2 (3 wt%)' electrolytes (inset of figure).

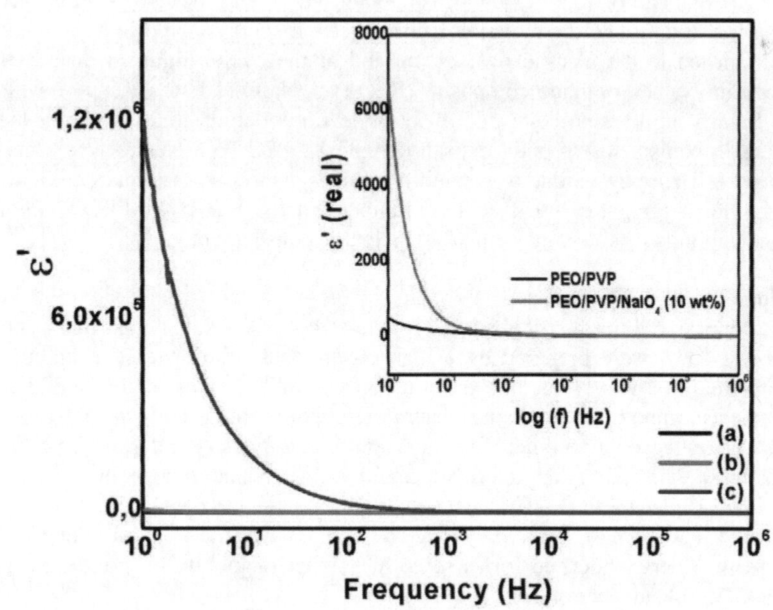

Figure 4. ε' vs frequency plots for 'PEO/PVP/NaIO_4(10 wt%)/TiO_2(3 wt%)' (blue line) and PEO/PVP (black line), salt complexed PEO/PVP (red line) electrolytes (inset of figure).

The ionic conductivity of the samples is calculated by the equation, $\sigma = t/(R_b*A)$, where t and A represent thickness and area of the electroldes, respectively. The bulk resistance is obtained from the intercept of the semicircle at the high frequency side (1 MHz to 1 Hz) of the plot, with the real axis (as shown in figure 3).The estimated room temperature conductivity of pure PEO/PVP blend electrolyte is 2.24 X 10^{-9} S/cm and associated ionic conductivity for 'PEO/PVP/$NaIO_4$ (10 wt%)' electrolyte increased to 1.57 X 10^{-7} S/cm. As a result of inclusion of nano-sized TiO_2 (3 wt%) of filler, the intercept of higher end of semicircle corresponding to 'PEO/PVP/$NaIO_4$ (10 wt%) electrolyte' on the Z^{I} - axis tends to move towards lower value as shown in inset of figure 3. The corresponding room temperature conductivity of nan-composite electrolyte is increased to 9.82 X 10^{-6} S/cm. The enhancement of conductivity as a result of inclusion of TiO_2 nano fillers could be due to reduction in crystallinity of the polymer chains in blend polymer electrolyte and improve the capacity of sodium ion transport in the electrolyte film and hence improve the ionic conductivity of the polymer electrolyte film [42, 43]. Dielectric materials are recognized as a media that have an ability to store electrical energy. This property of dielectric materials can be estimate by measuring is the permittivity or di-electric constant of the material. In alternating electric fields the relative permittivity exhibits complex behavior and is defined by $\varepsilon^* = \varepsilon' - j\varepsilon''$. The real ($Z_r$) and imaginary ($Z_i$) parts of complex impedance (Z^*) are also used for the evaluation of real and imaginary parts of dielectric permittivity using the following equations [44]

$$\varepsilon' = \frac{Z_i}{\omega C_0(Z_r^2+Z_i^2)} \quad (1)$$

$$\varepsilon'' = \frac{Z_r}{\omega C_0(Z_r^2+Z_i^2)} \quad (2)$$

Here C_o is the vacuum capacitance and given by $\varepsilon_o A/d$, where ε_o is a permittivity of free space and is equal to 8.85×10^{-12} Fm^{-1}. The angular frequency is given as $\omega = 2\pi f$, where f is the frequency of applied field. The real part of complex dielectric permittivity ε' has the same significance as that of the ordinary dielectric constant of the material. It measures the energy stored in the material during each cycle, to be returned to the electric field at the end of the cycle. Figure 4 describes the variation dielectric constant values of nano-composite 'PEO/PVP/$NaIO_4$(10 wt%)/TiO_2(3 wt%)' (blue line) electrolyte. Significantly, nano-composite electrolyte films demonstrate relatively higher dielectric constant values in comparison to pure and salt complexed PEO/PVP electrolytes films (as shown in inset of figure 4). This may be due to capability of TiO_2 nano particles leads to dissociation of un-dissociated salt/ion aggregates into free ions (anions) in the matrix of PEO/PVP/$NaIO_4$(10 wt%) electrolyte and supports for the enhancement of dielectric constant values [45].

4. Conclusions

In the present report, $NaIO_4$ salt complexed and nano-sized TiO_2 fillers doped free standing PEO/PVP blend electrolyte films were prepared by using conventional solution cast technique. XRD studies revealed the increase of amorphous portion in the matrix of TiO_2 nano composite blend electrolytes. Increase of surface smoothness of blend electrolytes upon doping TiO_2 nano fillers confirmed the enhancement of degree of amorphicity in the blend electrolytes. Relatively, TiO_2 nano composite blend electrolytes showed good thermal stability. The nano-composite 'PEO/PVP/$NaIO_4$(10 wt%)/TiO_2(3 wt%)' electrolyte demonstrated higher room temperature ionic conductivity of 9.82 X 10^{-6} S/cm. The increase in the dielectric constants as a result of addition of 10 wt% of salt and 3 wt% TiO_2 nano fillers evidenced for increase in number of mobile ions and their mobility in the matrix of PEO/PVP blend electrolytes.

Acknowledgments:
The authors gratefully acknowledges the INERA EU project Research Potential "Research and Innovation Capacity Strengthening of ISSP-BAS in Multifunctional Nanostructures" (FP7 REGPOT-

2012-2013-1) support for his participation at the INERA Workshop "Membrane and Liquid Crystal Nanostructures" (MELINA 2016), 3-6 September, 2016, Varna, Bulgaria.

References

[1] Boschin A and Johansson P 2015 *Electrochimica Acta* **175** 124
[2] Yoshino A 2012 *Angew Chemie-International Ed.* **51** 5798
[3] Zhou G, Li F F and Cheng H M 2014 *Energy Environ. Sci.,* **7** 1307
[4] Vikström H, Davidsson S and Hook M 2013 *Appl. Energy* **110** 252
[5] Palomares V, Serras P, Villaluenga I, Hueso K B, Carretero-Gonzalez J and Rojo T 2012 *Energy Environ Sci.* **5** 5884
[6] Vignarooban K, Kushagra R, Elango A, Badami P, Mellander B.-E, Xu X, Tucker T G, Nam C and Kannan A M 2016 *International Journal of Hydrogen Energy* **41** 2829
[7] Xu J, Dou S, Liu H and Dai L 2013 *Nano Energy* **2** 439
[8] Lu L, Han X, Li J, Hua J and Ouyang M 2013 *J. Power Sources* **226** 272
[9] Ni'mah Y L, Cheng M-Y, Cheng J H, Rick J and Hwang B-J 2015 *J. Power Sources* **278** 375
[10] Wang L, Lu Y, Liu J, Xu M, Cheng J, Zhang D, Goodenough J B 2013 *Angew. Chem.* **52** 1964
[11] Patel M, Chandrappa K G and Bhattacharyya A J 2010 *Solid State Ionics* **181** 844
[12] Mohan V M, Raja V, Bhargav P B, Sharma A K and Rao V V R N 2007 *J. Polym. Res.* **14** 283
[13] Mohapatra S, Thakur A and Choudhary R N P 2008 *Ionics* **14** 255
[14] Jimin S, Dong-Gyun K, Hee J K, Jin H L, Ji-Hoon B and Jong-Chan L 2014 *J. Mater. Chem. A* **2** 13873
[15] Armand M B 1986 *Annu. Rev. Mater. Sci.* **16** 245
[16] Tarascon J.-M and Armand M 2001 *Nature* **414** 359
[17] Armand M 1994 *Solid State Ionics* **69** 309
[18] Do C, Lunkenheimer P, Diddens D, Götz M, Weiß M, Loidl A, Sun X.-G, Allgaier J and Ohl M 2013 *Phys. Rev. Lett.* **111** 018301
[19] Laasonen K and Klein M L 1995 *J. Chem. Soc. Faraday Trans.* **91** 2633
[20] Wu H and Wick C D 2010 *Macromolecules* **43** 3502
[21] Neyertz S and Brown D 1996 *J. Chem. Phys.* **104** 3797
[22] Balbuena P B 2005 *Electrochim. Acta* **50** 3788
[23] Neyertz S, Brown D and Thomas J O 1995 *Electrochim. Acta* **40** 2063
[24] Aabloo A and Thomas J 2001 *Solid State Ionics* **143** 83
[25] Hektor A, Klintenberg M K, Aabloo A and Thomas J O 2003 *J. Mater. Chem.* **13** 214
[26] Kumar K K, Ravi M, Pavani Y, Bhavani S, Sharma A K and V.V.R. Narasimha Rao 2011 *Phys. B: Condens. Matter* **406** 1706
[27] Stephan A M and Nahm K S 2006 *Polymer* **47** 5952
[28] Kumar K K, Ravi M, Pavani Y, Bhavani S, Sharma A K and Rao V V R N 2014 *Journal of Membrane Science* **454** 200
[29] Inganas O 1998 *J. Br. Polym.* **20** 223
[30] Kim D W, Park J K and Rhee H W 1996 *Solid State Ion.* **83** 49
[31] Di Marco G, Lanza M and Pieruccini M 1996 *Solid State Ion.* 89 117
[32] Wieczorek W and Stevens J R 1997 *J. Phys. Chem.* **101 (9)** 1529
[33] Borghini M C, Mastragostino M, Passerini S and Scrosati B 1995 *J. Electrochem. Soc.* **142** 2118
[34] Kumar D and Hashmi S A 2010 *Solid State Ionics* **181** 416
[35] Bhide A and Hariharan K 2008 *Polym. Int.* **57** 523
[36] Agrawal R C, Hashmi S A 2007 *Ionics* **13** 295

[37] Arup Dey S, Karan S and De S K 2009 *Solid State Commun.* **149** 1282
[38] Noor S A M, Ahmad A, Talib I A and Rahman M Y A 2010 *Ionics* **16** 161
[39] Chu P P and Reddy M J 2003 *J. Power Sources* **115** 288
[40] Zang S, Lee J Y and Hong L 2004 *J. Power Sources* **134** 95
[41] Chun-Chen Y 2002 *J. Power Sources* **109** 22
[42] Lin C W, Hung C L, Venkateswarlu M, Hwang B J 2005 *J. Power Sources* **146** 397
[43] Lin Y -C, Cheng J -H, Venkateswarlu M, Wang F -M, Santhanam R, Hwang B J 2012 *J. Chin. Chem. Soc.* **59** 1250
[44] Aziz S B, Abidin Z H Z and Arof A K 2010 *Express Polym. Lett.* **4** 300

IOP Conf. Series: Journal of Physics: Conf. Series **780** (2017) 012007 doi:10.1088/1742-6596/780/1/012007

Electro-optical and dielectric characterization of submicrometer-sized PDLC films

Hari Krishna Koduru [1,2]
[1] Georgi Nadjakov Institute of Solid State Physics, Bulgarian Academy of Sciences, 72 Tzarigradsko chaussee, Sofia 1784, Bulgaria
[2] Dipartimento di Fisica, Università degli Studi della Calabria, Via P. Bucci, Cubo 31C - 87036 Rende (CS), Italy

E-mail: koduruharika@gmail.com

Abstract. This report concerns thin films (25 μm thickness) of polymer-dispersed liquid crystal (PDLC) composites having submicrometer-sized nematic liquid crystal domains. Such PDLC films were prepared by performing photo-induced phase separation of mixed soft material composed from the room-temperature nematic liquid crystal 4-*n*-heptyl cyanobiphenyl (7CB) and photo-curable polymer NOA-65. Being of interest for electro-optical (EO) and other applications driven by alternating-current electric field, the produced composites were studied by EO measurements and complex impedance spectroscopy. For the examined soft-solid composite material with a large content of submicro-confined nematic, was obtained information about the dielectric permittivity and molecular dynamics, useful for practice.

1. Introduction

The Polymer-Dispersed Liquid Crystals (PDLCs) [1-3] represent an important class of materials with useful electro-optical (EO) applications (without the need for optical polarizers) spread in a variety of areas such as flexible displays, integrated optics, sensorics, smart windows, etc. [4-17]. The PDLCs are two-phase systems containing liquid crystal (LC) domains, usually in the forms of droplets dispersed into a solid optically-transparent polymer matrix. The effective refractive index of the dispersed LC is mismatched with the index of the polymer matrix, which causes anisotropic multiple light scattering in the PDLCs. As known, the PDLC devices operate through the change in the effective refractive index of the dispersed LC material by applying alternating current (AC) electric field, thus achieving the EO switching between a largely scattering (OFF) and optically transparent (ON) state. Macroscopically, the operating mechanism of PDLCs is the electrically-induced dielectric reorientation, as described in details in the literature [1,18]. Electric fields of a sufficient strength may change the orientational direction of the mesogenes in the LC droplets. Thus, the optical properties of PDLCs can be electrically-commanded.

Especially, the sub-micrometer PDLCs and PDLCs on nanometer scale (often known also as 'nano-PDLCs') were achieved a wide attention for producing smart EO materials and for devices for active control of light, e.g., beam steering/splitting/multiplexing, fast-responsive amplitude-frequency and phase modulators, as well as fast and polarization-independent optical shutters (or intensity modulators) for, e.g., telecommunication systems [19-24]. Also, such PDLC systems have been applied in electrically-commanded diffractive optics, e.g., for Fresnel lens, optical phase gratings, grating beam splitters, etc. [25,26].

When one considers the EO response of PDLCs, their dielectric properties should be also taken into account. Numerous studies of the dielectric properties (and their frequency characteristics) of PDLC films have been reported (e.g., [27-34]. Also flexo-dielectric spectroscopy was successfully applied for investigation of flexoelectric and dielectric response of PDLCs [10,35-38]. In the work presented here, thin films (25 μm thickness) of submicrometer-size PDLCs produced from the room-temperature nematic LC 4-*n*-heptyl cyanobiphenyl (commercially known as 7CB) dispersed in a transparent matrix of the photo-curable polymer NOA65, were experimentally studied with EO measurements and dielectric spectroscopy. By means of the latter technique, previously applied for investigation of soft-solid nanocomposite systems [39,40], the dielectric characteristics of the examined 7CB/NOA65 PDLC were obtained within the frequency range from 1 mHz to 100 kHz of the driving electric field.

2. Experimental

The PDLCs studied here were produced by ultraviolet (UV)-photopolymerization-induced phase separation (PIPS). By this general method, PDLC structures were formed initiated from a mixture of reactive photo-curable monomer and LC molecules. The PDLCs films with a size of 1 cm × 1 cm and a thickness of 25 μm were formed in cells assembled from two 1 mm-thick glass plates, to serve as electrodes, each coated inside by ultrathin transparent conductive layers of indium tin oxide (ITO). The ITO-glass substrates were not treated and had no rubbing. At room temperature, the LC 4-n-heptyl cyanobiphenyl (7CB) has a stable nematic phase. The nematic-to-isotropic transition temperature (the clearing point) of 7CB is ca. 42 °C. Identical cells were filled with pure 7CB LC for reference measurements.

The PDLC composite structure was created by irradiation of a homogeneous mixture of 7CB (from Merck) and the liquid pre-polymer NOA65 (Norland Optical Adhesive, from Norland Products Inc., New Brunswick, N.J.) in a weight ratio 50 % : 50 %. A uniform UV illumination at the wavelength of 365 nm from continuous Hg-lamp was employed. The UV light intensity was ~ 37 ± 3 mW/cm^2. The UV curing time was ~ 15 min. The transmittance of the ITO layers at the wavelength of 365 nm was 81.5 %, as measured by photometry. The refractive index of the cured NOA65 polymer is n_p = 1.524 (at λ = 633 nm and 20 °C). The nematic 7CB is characterized by the following refractive indices at 25° C and λ = 633 nm: n_o (ordinary refractive index) = 1.517, and n_e (extra-ordinary refractive index) = 1.713 [41]. Therefore, the LC content of the PDLC is not only highly locally birefringent ($\Delta n = n_e - n_o$ ~ 0.2), but there is also a good index match between no and np.

The morphology of the prepared 7CB/NOA65 PDLC films was viewed through optical microscope NU-2 Universal Research Microscope (Zeiss) and recorded by Hitachi VK-C150ED video camera and computer. The EO response of the films upon an alternating-current (AC) electric field was investigated by use of circularly polarized He-Ne laser (λ = 632.8 nm, 1 mW) in the normal incidence direction on the PDLC sample. The intensity of He-Ne laser beam transmitted through the PDLC films was measured with a photodiode and lock-in technique (SR830 DSP lock-in amplifier, Stanford Research Systems). The measurement of electrical impedance of the PDLC was performed by Potentiostat/Galvanostat SP-200 (product of Bio-Logic). The complex impedance spectra in the frequency range f = 0.001 – 10^5 Hz were recorded at a probe voltage of 0.5 V_{RMS} in the sinusoidal waveform. All measurements were done at room temperature.

3. Results and Discussion

The micrographs in figure 1 show the formed morphology of a fabricated submicrometer-PDLC film. Closely inspected, one can see that to some extent the LC dispersed within the polymer exists as domains separated by polymer networks rather than isolated LC droplets. This kind of morphology is well known for the microsized PDLCs [1]. The situation is similar to that reported in [42] for microscale PDLC system. As seen from figure 1, the PDLC composite contains mostly very small LC domains whose characteristic size can be estimated to be less than 1 μm. Some isolated droplet-like creatures as large as 2 μm can be also observed. The 7CB domains captured within the solid NOA65 polymer were randomly dispersed in all directions and their size distribution was not uniform.

Actually, the examined PDLC films obtained by PIPS exhibit phase-separated LC domains varying from nano to micrometer sizes. Thus, this material of multi-sized LC domains considerably differs from UV-light-cured monodisperse PDLCs that exhibit a narrow distribution of the LC droplet size (e.g., [43,44]), as well as from high-power pulsed UV laser-cured single-layered PDLC films with a gradient droplet-size distribution [45-49]. Note that for the preparation of authentic nano-PDLC composite material, the polymer concentration must be higher (e.g., > ~70 wt% [24]) than the one (50 wt%) used here.

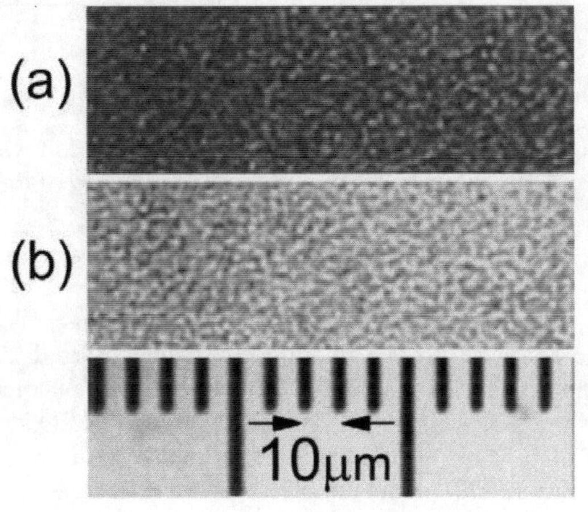

Figure 1. Optical microscopy images of the studied submicrometer-PDLC as taken with no polarizers. The applied voltage was zero (a) and 60 V_{RMS} (b). The temperature of the PDLC film was 26 °C.

The large spread of the size distribution of the LC domains should be reflected in the EO switching behavior of PDLC. As seen from the voltage-dependent light transmittance (figure 2), the studied PDLC is switched by applying an external AC electric field, between its scatter-translucent (OFF) and transparent (ON) states, in fact. This occurs due to the positive dielectric anisotropy of the LC 7CB ($\Delta\varepsilon$ ~ 10 at a temperature of 25 °C and 1 kHz frequency [50]), thereby the driving electric field forces the nematic director reorientation towards the field direction (that coincides with the direction of the light propagation, as well). As a result, the refractive index mismatch between the LC domains and the polymer matrix is removed – the PDLC becomes transparent (in its ON state). Taking into account the transmittance and reflectance of the ITO glasses, the controllable transmission range between the ON and OFF states of the PDLC was evaluated to be about 75 % in the visible range.

It is seen from figure 2(a) that the examined 7CB/NOA65 PDLC exhibited no distinct threshold of EO activation. The threshold voltage (V_{th}) could be estimated to be ~ 30 V_{RMS} (see figure 2 b), i.e. threshold electric field E_{th} = 1.2 V/μm – a value that is less than E_{th} of typical UV-light-cured micro-scale PDLCs [1,18,51], but is comparable with E_{th} for monodisperse micrometer-sized PDLC systems [16,43,44]. Reasonably, E_{th} of present 7CB/NOA65 PDLC is considerably larger than E_{th} for single-layered microsized PDLC with relatively large LC droplets, where E_{th} can be ~ 0.1 V/μm, or even less [38,49,52]. For comparison, the required switching voltages for typical nanoscale PDLCs are well greater than 30 V_{RMS}, and the saturation voltage (for ON-state) is typically ~ 10 V/μm [22,25,26], i.e. for a 25 μm cell gap, the required operating voltage can reach ~ 250 V_{RMS}. Thus, the voltage range of EO switching of the present hybrid (multiscale) PDLC merges the switching voltages of micro-scale and nano-scale PDLCs. Note, due to the presence of micro-scale LC droplets, the contrast ratio (4.4) of the hybrid PDLC film under study is significantly higher than the contrast of the usual nano-PDLCs at the normal direction [19-26].

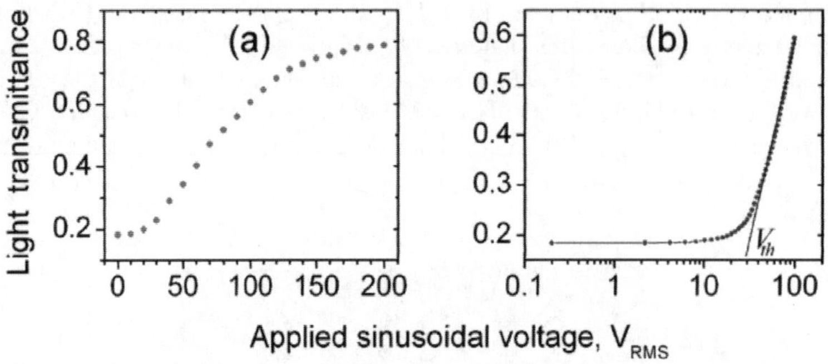

Figure 2. Voltage-dependent light transmittance measured for the cell with submicrometer-sized PDLC. Voltage increment: 10 V_{RMS} (a) and 2 V_{RMS} (b). The frequency of the driving electric field was 1 kHz, the temperature of the cell was 25° C.

Reasonably, the large voltage interval between V_{th} and the voltage relevant to the ON state (see figure 2 a) results from the presence of large amount of micrometer-sized LC droplets (responsible for the low-voltage EO switching) besides the confined LC domains of nanometer sizes (responsible for the high-voltage EO switching, respectively). The extended range of EO switching voltage makes the studied 7CB/NOA65 PDLC not well suited for efficient electrically-controllable optical switches/shutters and amplitude EO modulators. However, such PDLC films can still be used as switchable attenuators for control of light, as well as for simple regulators/discriminators of light intensity. It should be noted that PDLC films under study have the property of polarization independence, in the normal incidence direction. On the other hand, the morphology of such multiscale PDLC films is not proper for phase-controllable, diffractive and adaptive optics. Nevertheless, the produced PDLC composites may be of interest for other EO applications, or as a soft-solid dielectric material. That is why, the submicron-sized 7CB/NOA65 PDLC was characterized here by dielectric spectroscopy, as follows below.

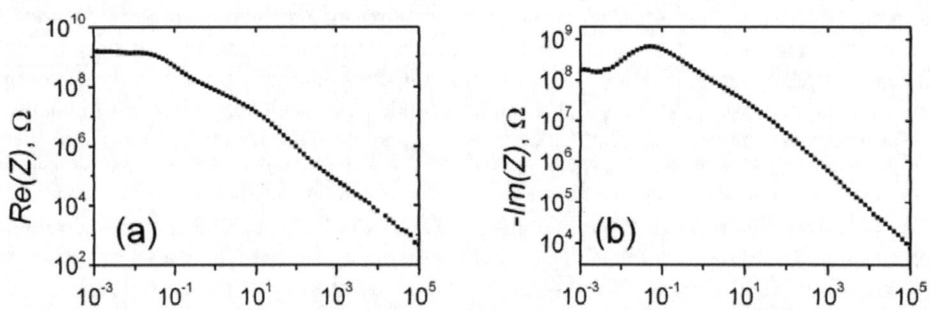

Figure 3. Raw data obtained by impedance spectroscopy: (a) real and (b) imaginary parts of complex impedance measured for the studied submicron-sized 7CB/NOA65 PDLC film.

Data obtained by impedance spectroscopy are shown in figure 3. From the recorded complex impedance spectra (the real (ReZ) and imaginary (*Im*Z) parts of impedance as a function of the frequency of the applied electric field), the frequency-dependent complex dielectric function $\varepsilon^* = \varepsilon' - i\,\varepsilon''$ can be obtained by a procedure, for example, used in Ref. [53]. The calculated dielectric spectra of the studied PDLC are shown in figure 4. To distinguish the contribution of both components of PDLC (the LC and the polymer) in the dielectric behavior, the modulus $|\varepsilon^*|$ of the complex permittivity as a function of the electric field frequency was fitted by applying the approach described in Ref. [53]. Thus, the fit of the dielectric spectrum for the considered PDLC was performed by use of Kohlrausch-Williams-Watts (KWW) [54,55] and Havriliak-Negami (HN) [56,57] functions.

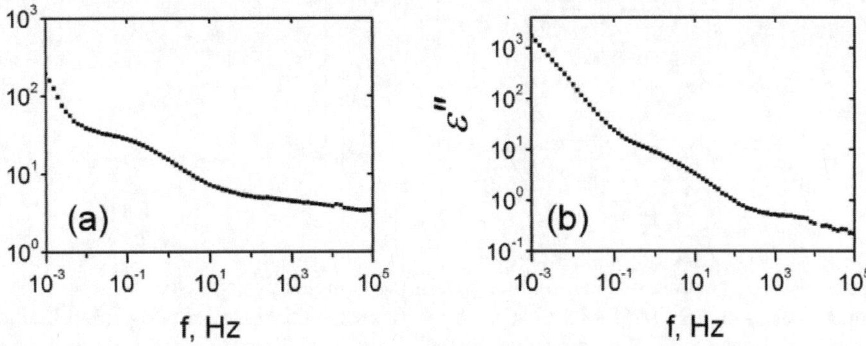

Figure 4. Dielectric spectra calculated for submicron-sized 7CB/NOA65 PDLC: real (a) and imaginary (b) parts of complex dielectric function calculated from data in figure 3.

Basically, the KWW model is successfully applied for description of polymeric materials, whereas the HN relaxation function is excellent for modeling of LCs. Thus, the fit on the obtained data of the complex dielectric permittivity was done using a combined model that contains a term for the DC conductivity, and, because the system is composed by a polymer and a LC, has been used a term with the KWW-model, to take account the response of the polymer, and a term with the HN-model, to take account for the LC response. The coupling between the universal HN form for the dielectric function (the third term) and more general KWW function for time-domain relaxation (the second term), with the addition of a conductivity contribution present at lower frequencies (the first term) can be expressed as:

$$\varepsilon^* = \frac{\sigma}{\varepsilon_0 (i2\pi f)^n} + F.T.\left\{ \Delta\varepsilon_{KWW} \exp\left[\frac{-t}{\tau_{KWW}}\right]^{\beta} \right\} + \frac{\Delta\varepsilon_{HN}}{\left[1 + i\, f/(f_0)^a\right]^b} \qquad (1)$$

In Eq. (1), f [Hz] is the frequency of the applied electric field, σ [S/cm] is the DC electrical conductivity of the system and takes into account the free charges diffusion, f_0 [Hz] is the relaxation frequency, ε_0 is the vacuum permittivity, $\Delta\varepsilon$ is the dielectric strength, n is the power factor of the conductivity term. The fit parameters a and b are shape parameters accounting for the asymmetry and broadness of the dielectric dispersion curve (power indexes that define the low- and high-frequency limits of ε'', respectively). F.T. means the Fourier transform, β is the stretching exponent, τ_{KWW} [s] is the time for dielectric relaxation, and t is the time variable. The fits to the measured dielectric data were performed using a least-squared routine.

The result is given in figure 5(a). The $|\varepsilon^*|$ frequency spectrum of the PDLCs can be considered as consisting (convolution) of two main peaks associated with the contributions of both the LC and the amorphous polymer. As such, by the last two terms in the expression of the fitting function in Eq. (1) one can decompose the $|\varepsilon^*|$ spectrum into two peaks, resulting from the KWW function and HN function, respectively. The two resolved peaks of dielectric relaxation response can be seen in figure 5(a). Table 1 summarizes the values of the fitting parameters obtained for each individual part of the fit, as well as for the total fit. The quality of the fit values was excellent. In fact, the coefficient of determination for the fit obtained, was $R^2 = 0.99993$ (practically 1), and $\chi^2 = 0.00037$.

IOP Conf. Series: Journal of Physics: Conf. Series 780 (2017) 012007 doi:10.1088/1742-6596/780/1/012007

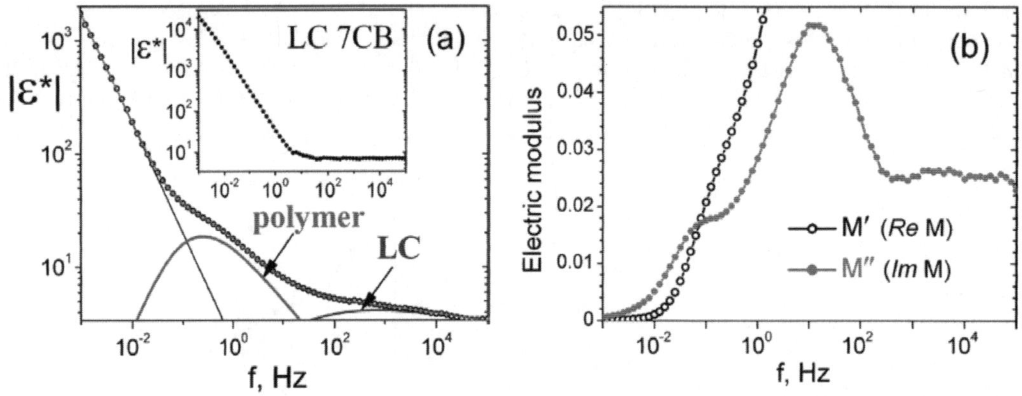

Figure 5. (a) The spectrum of the modulus ($|\varepsilon^*|$) of the complex permittivity (symbols) calculated for submicron-sized 7CB/NOA65 PDLC. The curves represent the resolved two peaks obtained by the KWW model (green) and HN model (blue), respectively, as well as their envelope (red), according to the best fit by the function in Eq. (1). The straight line represents the contribution of the conductivity term. The insert: the dielectric function of pure 7CB LC, measured under identical experimental conditions. (b) Enlarged shapes of the calculated complex electric modulus of the PDLC under study.

Table 1. Best-fit parameters for the dielectric response of submicron-sized 7CB/NOA65 PDLC as modelled by Eq. (1).

Log(σ)	*n*	KWW: log(τ)	KWW: Δε	KWW: β	KWW: log(f_{max})
-11.95044	0.97422	-0.3027	59.28595	0.58015	-0.60715

HN: log(f_o)	HN: Δε	HN: a	HN: b	HN: log(f_{max})
1.23492	54.00488	0.44279	0.18088	2.85651

Further, the real (M′) and imaginary (M″) parts of the electric modulus of the studied PDLC can also be analyzed. These quantities are expressed as: $M' = \varepsilon'/(\varepsilon'^2 + \varepsilon''^2)$ and $M'' = \varepsilon''/(\varepsilon'^2 + \varepsilon''^2)$. For the frequency dependence of M″ one can clearly see two peaks for the two materials that compose the PDLC (figure 5 b), corresponding with the values obtained by the KWW model (for the polymer) and HN model (for the LC), respectively.

In general, the dielectric data provide information about both (i) the electrical conduction behavior of extrinsic ions and (ii) the reorientation motions of dipoles. (i) By PDLCs, the application of the method used here does evidence the conductivity-related low-frequency process (mainly interfacial polarization due to the difference in conductivities between the droplets and polymer matrix). Conductivity data can be extracted from the slope of the straight line in figure 5(a). The value of the DC conductivity is typical for these materials. (ii) The dielectric behaviors can be used to obtain information on dipole relaxation processes in PDLCs. The dielectric behavior for the PDLC can be expected to be intermediate between that of the LC phase and the amorphous polymer matrix [30-32,34]. However, the overall dielectric relaxation in PDLCs is rather complex process related to contributions of various processes, e.g. the dipole motions in the LC and in the host polymer matrix [58]. In any case, the dielectric spectroscopy is a convenient non-optical method for determining the alignment of the LC phase of a PDLC when AC field is applied. Upon AC voltage, the LCs and

PDLCs switch optically through the alignment of the LC director along the driving electric field, and this gives marked changes in their dielectric properties.

4. Conclusion

In summary, in this report are presented experimental results for 7CB/NOA65 PDLC with a broad size-distribution of the LC domains ranging from nanoscale to microscale, but predominantly submicron-sized. The electro-optical and dielectric response of the studied PDLC thin films was characterized at room temperature. The dielectric behavior of the PDLC with submicrometer-LC-domains was rationalized in terms of a model equivalent that included its components: the LC phase, an amorphous polymer matrix containing dispersed LC material, and an interlayer region in which the LC material is immobile. For the examined PDLC, an analysis of the dielectric data with a two-phase model gave a satisfactory account of the experimental behavior. If investigated the changes in the dielectric behavior of the PDLC upon a driving electric field, the additional information obtained about the complex permittivity and molecular dynamics of this soft-solid composite material containing submicro-confined nematic should be useful for stabilizing and optimizing its EO function.

Acknowledgments

The author gratefully acknowledges the INERA EU project Research Potential "Research and Innovation Capacity Strengthening of ISSP-BAS in Multifunctional Nanostructures" (FP7 REGPOT-2012-2013-1) support for his participation at the INERA Workshop “Membrane and Liquid Crystal Nanostructures” (MELINA 2016), 3-6 September, 2016, Varna, Bulgaria.

References

[1] Drzaic P S 1995 *Liquid Crystal Dispersions* (Singapore: World Scientific Publishing)
[2] Doane J W 2006 *Liq Cryst* **33** 1313
[3] Drzaic P 2006 *Liq Cryst* **33** 1281
[4] Wu S T and Yang D K 2001 *Reflective Liquid Crystal Displays* (New York: Wiley)
[5] Bloisi F and Vicari L 2003 Polymer-dispersed liquid crystals *Optical Applications of Liquid Crystals, Series in Optics and Optoelectronics* eds L Vicari (Bristol: Institute of Physics Publishing) chapter 4 p 148
[6] Wu S T and Fuh A Y G 2008 *Opt Commun* **281** 1732
[7] Malik P, Bubnov A M and Raina K K 2008 *Mol Cryst Liq Cryst* **494** 242
[8] Marinov Y G, Hadjichristov G B and Petrov A G 2009 *J Optoelectr Adv Mater* **11** 1186
[9] Park S and Hong J W 2009 *Thin Solid Films* **517** 3183
[10] Lallana P C, Vazquez C, Vinouze B, Heggarty K and Montero D S 2009 *Mol Cryst Liq Cryst* **502** 130
[11] Hadjichristov G B, Marinov Y G and Petrov A G 2011 *Appl Opt* **50** 2326
[12] Jeon Y J, Lee G H, Jang J E, Hwang K Y, Ahmad F, Jamil M, Lee J and Jung J E 2012 *Liq Cryst* **39** 1314
[13] Marinov Y G, Hadjichristov G B, Petrov A G, Marino S, Versace C and Scaramuzza N 2013 *J Appl Phys* **113** 064301
[14] Jeon S, Kim B, Ye H and Hong S K 2014 *Mol. Cryst Liq Cryst* **597** 159
[15] Marinov Y G, Hadjichristov G B, Petrov A G, Marino S, Versace C and Scaramuzza N 2013 *C R Acad Bulg Sci* **66** 819
[16] Song P, Gao Y, Wang F, Zhang L, Xie H, Yang Z and Yang H 2015 *Liq Cryst* **42** 390
[17] G B Hadjichristov, Y G Marinov and A G Petrov 2016 *AIP CP* **1722** 290007
[18] Crawford G R and Zumer S 1996 *Liquid Crystals in Complex Geometries Formed by Polymer and Porous Networks* (London: Taylor & Francis Ltd)
[19] Matsumoto S, Houlbert M, Hayashi T and Kubodera K 1996 *Appl Phys Lett* **69** 1044
[20] Bunning T J, Natarajan L V, Tondiglia V P, Dougherty G and Sutherland R L 1997 *J Polym Sci Part B: Polym. Phys.* **35** 2825

[21] Lucchetta D E, Karapinar R, Manni A and Simoni F 2002 *J Appl Phys* **91** 6060
[22] Ren H, Fan Y H, Lin Y H, Wu S T 2005 *Opt Commun* **247** 101
[23] Liu Y J and Sun X W 2008 *Adv OptoElectron* **2008** 684349
[24] Choi W K, Hou S L, Chen J Y, Su G D J, Li Y M 2015 *Mol Cryst Liq Cryst* **612** 232
[25] Ren H and Wu S T 2002 *Appl Phys Lett* **81** 3537
[26] Ren H, Fan Y H and Wu S T 2003 *Appl Phys Lett* **83** 1515
[27] Miyamoto A, Kikuchi H, Kobayashi S, Morimura G and Kajiyama T 1991 *Macromolecules* **24** 3915
[28] Rout D K and Jain S C 1992 *Mol Cryst Liq Cryst* **210** 75
[29] Zhong Z Z, Schuele D E, Gordon W L, Adamic K J and Akins R B 1992 *J Polym Sci Part B: Polym Phys* **30** 1443
[30] Williams G, Shinton S E and Aldridge G A 2001 J Polym Sci B Polym Phys 39 1173
[31] Brás A R E, Viciosa M T, Rodrigues C M, Dias C J and Dionísio M 2006 Phys Rev E 73 061709
[32] Ganesan L M, Frübing P, Mellinger A and Gerhard R 2009 *J Phys D: Appl Phys* **42** 092006
[33] Krasinski F, Splingart B, Dubois F, Maschke U and Legrand C 2011 *Mol Cryst Liq Cryst* **542** 227
[34] Deshmukh R R 2015 Electro-optic and Dielectric Responses in PDLC Composite Systems *Liquid Crystalline Polymers* vol **2**-*Processing and Applications* ed V K Thakur and M R Kessler (Cham, Switzerland: Springer) chapter 7 p 169
[35] Hadjichristov G B, Pavlov L, Marinov Y and Petrov A G 2008 *Proc SPIE* **7027** 70271M
[36] Hadjichristov G B, Marinov Y and Petrov A G 2010 *Mol Cryst Liq Cryst* **525** 128
[37] Marinov Y G, Hadjichristov G B and Petrov A G 2010 *Opt Las Eng* **48** 1161
[38] Marinov Y G, Hadjichristov G B, Marino S, Todorova L, D'Elia S, Versace C, Scaramuzza N and Petrov A G 2012 *Bulg J Phys* **39** 92
[39] Infusino M, Luca A D, Ciuchi F, Ionescu A, Scaramuzza N and Strangi G 2014 *J Mater Sci* **49** 1805
[40] Marino L, Marino S, Wang D, Bruno E and Scaramuzza N 2014 *Soft Matter* **10** 3842
[41] Dunmur D A, Manterfield M R, Miller W H and Dunleavy J K 1978 *Mol Cryst Liq Cryst* **45** 127
[42] Ren H, Lin Y H, Fan Y H and Wu S T 2005 *Appl Phys Lett* **86** 141110
[43] Park S I, Park N H and Suh K D 2002 *Liq Cryst* **29** 783
[44] Ryu J H, Lee S G and Suh K D 2004 *Liq Cryst* **31** 1587
[45] Hadjichristov G B, Angelov Ch and Petrov S P 2007 *AIP CP* **899** 625
[46] Marinov Y, Hadjichristov G B and Petrov A G 2007 *J Optoelectr Adv Mater* **9** 417
[47] Hadjichristov G B, Marinov Y G and Petrov A G 2009 *Opt Mater* **31** 1578
[48] Marinov Y G, Hadjichristov G B and Petrov A G 2009 *Cryst Res Technol* **44** 870
[49] Hadjichristov G B, Marinov Y G and Petrov A G 2010 *AIP Conf Proc* **CP1203** 182
[50] Ratna B R and Shashidhar R 1976 *Pramana - J Phys* **6** 278
[51] Doane J W 1990 *Polymer dispersed liquid crystal displays in Liquid Crystals: Applications and Uses* vol 1 ed B Bahadur (World Scientific) p 361
[52] Hadjichristov G B, Marinov Y G and Petrov A G 2009 *J Optoelectr Adv Mater* **11** 1190
[53] Alzari V, Nuvoli D, Sanna V, Caruso T, Marino S and Scaramuzza N 2016 *AIP Advances* **6** 035005
[54] Kohlrausch R 1854 *Ann Phys Chem (Poggendorff)* **91** 179
[55] Williams G and Watts D C 1970 *Transactions of the Faraday Society* **66** 80
[56] Havriliak S and Negami S 1966 *J Polym Sci C* **16** 99
[57] Havriliak S and Negami S 1967 *Polymer* **8** 161
[58] Williams G 1994 Dielectric Relaxation Behaviour of Liquid Crystals *Molecular Dynamics of Liquid Crystals, NATO ASI Series* vol **431** eds G R Luckhurst and C A Veracini (Dordrecht: Kluwer Academic) chapter 17 p 431

Photoresponsive azo-doped aerosil/7CB nematic liquid-crystalline nanocomposite films: the role of polyimide alignment layers of the films

Georgi B Hadjichristov[1,*] **and Yordan G Marinov**[2]
[1] Laboratory of Optics and Spectroscopy, Institute of Solid State Physics, Bulgarian Academy of Sciences, 72 Tzarigradsko Chaussee, Sofia 1784, Bulgaria
[2] Laboratory of Biomolecular Layers, Institute of Solid State Physics, Bulgarian Academy of Sciences, 72 Tzarigradsko Chaussee, Sofia 1784, Bulgaria
E-mail: georgibh@issp.bas.bg

Abstract. We studied thin films (25 μm-thick) of nanomaterials composed from 3 wt.% aerosil nanospheres and the room-temperature nematic liquid crystal 4-*n*-heptyl cyanobiphenyl (7CB). The inclusion of 3 wt.% of the photoactive liquid crystal 4-(4'-ethoxyphenylazo)phenyl hexanoate (EPH) in the aerosil/7CB nanostructured nematics make them photoresponsive. The films had alignment layers from rubbed polyimide (PI). Our study is concentrated on the inspection of the PI-role for the photo-stimulated electro-optical properties of the considered EPH-doped aerosil/7CB nanocomposite films.

1. Introduction

Relatively recently, the research on the liquid-crystal (LC)-silica composites has demonstrated that such materials have a potential for a number of optical, electro-optical (EO) and electric applications [1]. The mechanism of light scattering of these nanostructured materals has been well established (see e.g., [2]). In particular, LC-aerosil systems have been extensively investigated [3-8]. The aerosil nanoparticles are silica nanoscale spheres whose surfaces are decorated to achieve hydrophilic or hydrophobic interactions.

Recently, we have studied the scatter-based electro-optics of nanocomposites formed from 3 wt.% hydrophilic aerosil nanospheres (~ 7 nm size) within the nematic LC 4-*n*-heptyl cyanobiphenyl (7CB) [**9**]. Further, the ultraviolet (UV) light-modified EO response of the aerosil/7CB nanofilled nematic doped with 4-(4'-ethoxyphenylazo)phenyl hexanoate (EPH) azo-molecules was reported [10]. In the work presented here, our aim is to be inspected the role of the rubbed polyimide (PI) alignment layers for the photo-stimulated EO response of EPH-doped aerosil/7CB films, in particular films with a thickness of 25 μm. The LC alignment by thin layers of a PI is a most common way for inducing a homogeneous (planar) alignment of the LC molecules at the interfaces, and the alignment effect of PI layers on the optical films is widely exploited for a long time in organic electronics and LC devices for use in photonics products and displays [11-13]. The rubbed PI alignment layers serve to align LC molecules in a certain orientation and conformation with a specific pretilt angle (the angle between the

* To whom any correspondence should be addressed.

director of the LC molecules and the PI alignment layer). To a great extent, the pretilt angle determines the EO performance of the LC devices.

We have focused our study on the voltage-dependent optical transmittance (the static EO behavior) of visible light passed through the EPH-doped aerosil/7CB nanocomposite films, being under AC electric field and exposed to illumination with continuous light sources. The transmittance of such films with alignment PI layers was compared to the corresponding characteristics of the same nanocomposite films (either EPH-doped or undoped) but without PI layers, and the effect from the PI layers was elucidated by electrical and EO measurements.

2. Experimental

Nanospheres of hydrophilic Aerosil 300 having a diameter of ~ 7 nm and capped with silanol groups) at a concentration of 3 wt.% were dispersed in the well known room-temperature nematic LC 4'-heptyl-4-biphenylcarbonitrile (7CB). The preparation of $Aerosil_{300}$/7CB nanocomposite films has been described in details elsewhere [9]. The azo-containing LC compound 4-(4'-ethoxy phenylazo)phenyl hexanoate (EPH) [14] was added to the formed nanocomposite at a concentration of 3 wt.%. Thin films (with a thickness of 25 μm) of EPH-doped $Aerosil_{300}$/7CB gel were sandwiched between a pair of indium tin oxide (ITO) conducting glass plates. The ultrathin transparent ITO layers on the glass plates were overcoated with a thin layer of polyimide (PI) (namely, PI-2555 from HD Microsystems, USA) having unidirectional mechanical rubbing. The orientation directions of the two PI-layered substrates of the cells were antiparallel. Identical planar cells with PI layers overcoating the ITO glasses, but filled with undoped $Aerosil_{300}$/7CB, were also prepared in order to be used for reference measurements in our study. For the same purpose, the same two-component composite of undoped $Aerosil_{300}$/7CB, as well as the above mentioned $Aerosil_{300}$/7CB/EPH three-component nanocomposite were filled in cells with the same thickness of 25 μm but without PI alignment layers (i.e., with ITO glasses only).

The experimental set up has been described in details elsewhere [15]. Briefly, the EO response of EPH-doped $Aerosil_{300}$/7CB films to an alternating-current (AC) electric field was investigated by use of a circularly polarized He-Ne laser beam (wavelength λ = 632.8 nm, 1 mW power on the samples) and computer-controlled lock-in amplifier (SR830 DSP, Stanford Research Systems) coupled with a photodiode. The UV continuous light source was a LED (wavelength 375 nm). The UV light intensity on the samples was fixed at ~ 3 mW/cm^2 and kept the same during all the measurements in this work. The same applies to the light in the blue spectral range (wavelength 450 nm, from another LED whose intensity on the samples was ~ 30 mW/cm^2, sufficient to restore the initial state of the samples).

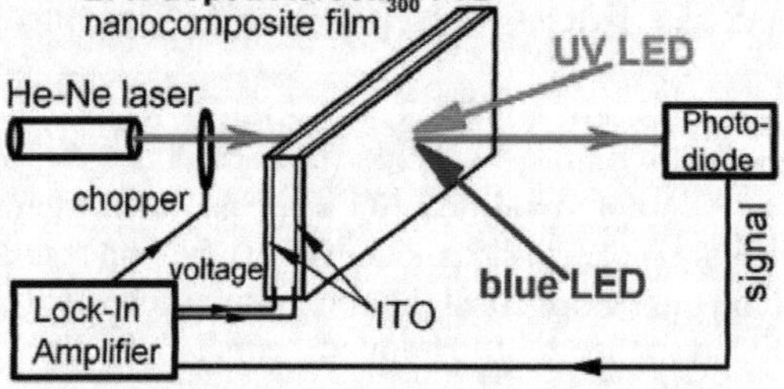

Figure 1. Schema of the optical setup used here.

The optical setup is schematically shown in figure 1. The voltage-dependent transmittance of EPH-doped Aerosil300/7CB nanocomposite films was measured at 1 kHz frequency of the electric field applied to the films and with He-Ne laser beam chopped at 90 Hz. No polarizers were used. By AC electric field transversally applied to the LC cells, the frequency dependent current through the nanocomposite films was measured by the lock-in instrument sweeping the frequency in the range from 1 Hz to 3 kHz at a fixed voltage of the electric field. During all the experiments, the temperature of the studied nanocomposite films was about 26° C.

3. Results and Discussion

Like the dispersion of 3 wt.% $Aerosil_{300}$ in 7CB [9, 16], the prepared 3 wt.% $Aerosil_{300}$/7CB doped with 3 wt.% EPH is also a soft gel nematic material. In the gel state of the filled nematics, an irregular network of hydrogen-bonded aerosil nanoparticles is formed that defines the geometrical confinement and the material properties, respectively. The inclusion of azo-bonded EPH mesogen makes possible to realize an efficient stimulation of the AC electric field-driven optical transmittance of EPH-doped $Aerosil_{300}$/7CB nanocomposite films [10]. In particular, this effect can be achieved by continuous illumination with relatively low-intensity UV light at the wavelength of 375 nm that coincides with the wavelength peak position of *trans-cis* absorption band of EPH compound [14]. In this case, the UV light-induced effect on the electro-optics of EPH-doped $Aerosil_{300}$/7CB occurs through *trans*-to-*cis* photoisomerization of EPH azo-bonded molecules (the transformation from rod-shaped *trans* isomers to bent-shaped *cis* isomers) [14,17-19]. The photoisomerization is reversible by employing continuous blue-light illumination. The latter leads to *cis*-to-*trans* photoisomerization of EPH that restores the initial state of the azo-molecules, and thereby the initial state of the EPH-doped $Aerosil_{300}$/7CB nematic system. The common mechanism of *trans-cis-trans* conversion by azobenzene molecules is well known, e.g., [20-22].

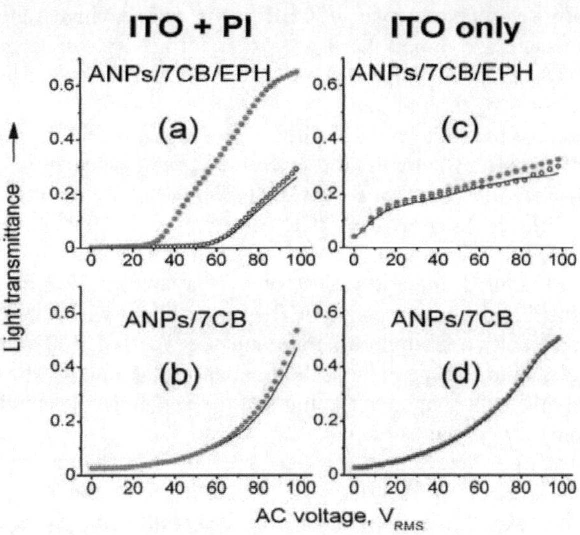

Figure 2. Voltage-dependent transmittance of: (a) EPH-doped $Aerosil_{300}$/7CB and (b) net $Aerosil_{300}$/7CB, both in identical cells with PI and ITO layers. The second couple: (c) EPH-doped $Aerosil_{300}$/7CB and (d) net $Aerosil_{300}$/7CB, both in identical cells without PI alignment layers (ITO glasses only). Data obtained by use of He-Ne laser beam, under identical experimental conditions. No UV light (lines); by exposure on UV light (solid circles) and by subsequent illumination with blue light (open circles).

Figure 2 summarizes the voltage-dependent light transmittance curves measured for the four kinds of samples we have prepared. The light intensity changes were fully consistent with our observations by optical microscopy. The transmittance curves in figure 2 were obtained either without or with illumination of the films with UV light at the wavelength of 375 nm. For the sake of comparison, the voltage-dependent characteristics of all the examined films were recorded keeping a constant intensity of the UV light. Also in order to be compared, the behaviors were measured in a way that equally scales the levels of transmittance of the four samples.

The voltage-induced increase of the transparency of $Aerosil_{300}$/7CB films by increase of the applied AC electric field has been demonstrated and the electrically-controllable scatter-based mechanism of voltage-dependent light transmittance of this nematic gel material has been discussed in details in [9]. Basically, the same physical mechanism takes place for the $Aerosil_{300}$/7CB/EPH gel films. This mechanism is governed by the aerosil network formed within these nanofilled nematic systems, but it is strongly influenced by EPH azo-molecules that decorate the network. In particular, the EPH LC molecules can be bond with the network nanostructures, and thereby create larger structures. This reflects in the very distinct threshold voltage of the local Freedericksz transition observed for $Aerosil_{300}$/7CB/EPH (figure 2 a) in contrast to the smeared one for the net $Aerosil_{300}$/7CB (figure 2 b). Clearly, the large reduction of the threshold voltage of the local Freedericksz transition in $Aerosil_{300}$/7CB/EPH nematic films upon UV illumination ultimately occurs through the *cis*-isomers of the EPH molecules. The bent-shaped EPH *cis*-isomers formed by UV irradiation can additionally enhance the disorder in the photosensitive $Aerosil_{300}$/7CB/EPH nanocomposites studied here. The disorder is intrinsic to the non-photosensitive $Aerosil_{300}$/7CB dispersion [9,16], like similar aerosil-disordered LCs [1-5,23-25].

As seen from figure 2, a well pronounced UV light-produced stimulation (an increase of the film transparency at a given voltage, in fact a large shift of the voltage-transmittance behavior to the lower values) occurs only for EPH-doped $Aerosil_{300}$/7CB films in cells having PI alignment layers (figure 2 a). The contrast to the same azo-doped nanocomposites formed in counterparts of cells with no PI (figure 2 c) indicates the paramount importance of the PI alignment layers for the UV light-induced effect.

There are two impacts of the PI layers. The first is the LC alignment through the rubbing of the PI layers. In thin films of pure 7CB nematic, the orienting relief formed by the groove pattern on the surface of the parallel-rubbed PI layers of the LC cells should totally align the nematic 7CB. In spite of the parallel-rubbed PI layers, in $Aerosil_{300}$/7CB nanocomposite films the aerosil network prevents the alignment in volume, where, actually, a lot of pretilt variations are possible. By that, some alignment of this nanostructured nematic occurs only in a small region close to the PI surface. It should be noted that in EPH-doped $Aerosil_{300}$/7CB nematic films with parallel-rubbed PI layers the azo-bonded EPH LC molecules can support the alignment of the 7CB LC molecules. In this case, the role of the azo-molecules is to transpose the LC alignment in the bulk of the system structure. Upon UV light, most probably the bent-shaped *cis*-isomers of the azo-molecules modify the anchoring in the nanoscale vicinity of the network nanostructure.

The second impact of the PI layers is the generation of free charge carriers by the PI polymer subjected to UV light. The creation of net negative charges in the PI alignment layers of LCs (at the PI surface and inside in the bulk of LCs) by UV exposure, as well as the surface modification of the PI alignment layers by UV illumination (changes in the surface composition and in the pretilt angle of the PI surface), are known for a long time, e.g., [26,27]. For instance, a broadband UV irradiation an average power density of 4.5 mW/cm^2 (comparable to the intensity level of ~ 3 mW/cm^2 of the UV light we have presently employed here) has been used in [26] to be altered the pretilt angle of PI alignment layers for nematic LC, and thereby to change the interaction between the nematic LC and PI alignment layers (in particular, PI-2555). Indeed, the exposure of the UV light on the PI layer was found to change drastically the surface morphology and anisotropy due to photochemical dissociation, bond breaking and subsequent oxidation reactions [26,28].

In our experiment, the relative low UV-light intensity did not lead to irreversible surface modifications and direct photoalignment effect, but most probably is enough to (slightly) change the anchoring conditions close to the PI surfaces, as well as in the bulk of the nanostructured nematic system. The space charge generated via the PI layers that overcoat the ITO layers of the LC cell plates can assist the modification of LC anchoring conditions. Besides the diffusion of the photo-produced charges in the sample [29] and their subsequent migration or trapping in various pathways, the charges created from the PI layers can give rise of photo-induced surface polarization that can lead to some changes of the nematic LC anchoring, an effect reported in a number of studies on PI-LC interface subjected to UV light, where the alignment mechanism of LCs by PI-layers on various substrates and the nature of surface modification have been elucidated [12,27,30,31]. Actually, the light-induced charge will not only modify the PI–LC interaction, but may also increase the direct current (DC) offset voltage generated in a planar LC cell aligned with PI layers [32]. Upon AC electric field, the trapping sites generated during UV illumination could trap the carriers and generate an additional DC offset voltage shift. The static electricity by rubbing PI is well known (e.g., [33]).

To address the situation with possible electrical charging effects in the studied nanocomposite nematic gels, complementary electrical measurements were performed. By the examined $Aerosil_{300}$/7CB/EPH films having rubbed PI layers, the measurements evidence a sizable UV light-induced generation of charge (the results are given below). This effect is most efficient at DC or low-frequency AC field (e.g., below 100 Hz). Thus, under such conditions, an UV-induced weak effect of anchoring change due to UV-induced photo-electric effect provoked by the PI layers surely takes place for the $Aerosil_{300}$/7CB nanocomposites studied here (either EPH-doped or undoped). However, this UV-induced effect is much weaker (both at the surface and in the bulk) than the UV-light controlled LC alignment achieved by the change of LC anchoring conditions through the LC azobenzene molecules. The contribution of the latter effect is much more pronounced.

At this point comes the question about the UV light-induced effect on the AC field-driven light transmittance still observed (especially pronounced at higher AC voltage) in EPH-undoped $Aerosil_{300}$/7CB in LC cells having orienting PI layers (figure 2 b). In fact, this effect on the EO response was very weak in comparison with EPH-doped $Aerosil_{300}$/7CB (figure 2 a), but well reproducible, too. The optical microscopy displayed no sizable UV-produced morphology changes in this case, in contrast to photo-sensitized $Aerosil_{300}$/7CB/EPH in identical LC cells (with PI) at equal values of the applied AC voltage. The same applies to the same (non-sensitive to UV light at 375 nm) undoped $Aerosil_{300}$/7CB gel, but in LC cells with no PI layers. Note, that in the latter case no UV light-induced effect on the EO response of the films was detected (figure 2 d). In the context of the discussion above, the UV light-induced effect on the electro-optics of undoped $Aerosil_{300}$/7CB in LC cells having orienting PI layers (figure 2 b) can be ascribed to an affect from alignment of this nanostructured nematic in a region close to the PI surface, or can originate in the local field change owing to free charge carriers generated by the PI-coating of the LC walls when exposed to UV light.

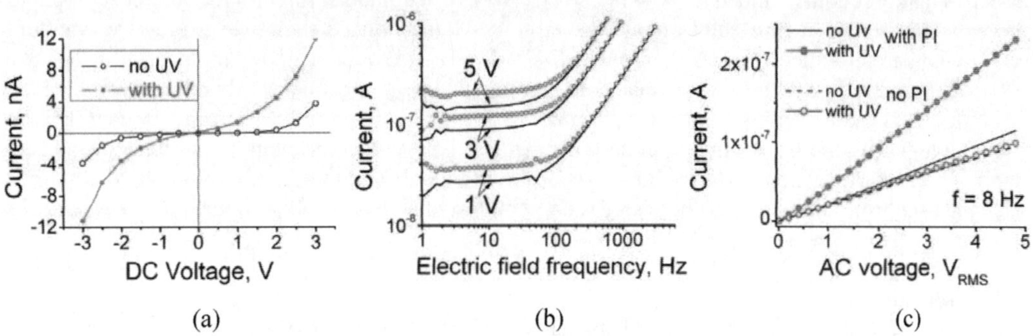

(a) (b) (c)

Figure 3. (a) DC current-voltage dependence measured for EPH-doped $Aerosil_{300}$/7CB nanocomposite film; (b) The frequency dependence of amplitude of the alternating current in EPH-doped $Aerosil_{300}$/7CB nanocomposite film. The spectra were obtained under identical experimental conditions. Pairs measured with no UV light (lines) and with UV light (circles), by applying AC voltages of 1 V, 3 V and 5 V (RMS); (c) AC current-voltage dependences measured for EPH-doped $Aerosil_{300}$/7CB in cells: (i) with PI coatings; (ii) with no PI coatings (ITO glasses only), under identical other experimental conditions. The frequency of the AC electric field applied on the cells was 8 Hz. Pairs measured with no UV light (lines) and with UV light (circles).

The generation of an excess free carriers due to UV illumination of the examined $Aerosil_{300}$/7CB/EPH nanocomposite films can be evidenced by electrical measurements upon static (DC) or dynamic (AC) field in the low-voltage regime. By that, a transversal electric-field configuration was used, through the ITO electrodes of the examined cells. The symmetric DC current-voltage (*I-V*) characteristics curve measured for the films (figure 3 a) displays a non-Ohmic behavior of increasing current by increasing voltage. Further, this behaviour gives an indication of very low leakage current and enhanced DC conductivity of the films being illuminated with UV light (under the experimental conditions of our measurements discussed above).

Figure 3(b) reports the frequency dependence of the alternating current by varying the strength of the AC electric field applied on the same EPH-doped $Aerosil_{300}$/7CB film under the same other experimental conditions. As seen, a well pronounced UV light-induced effect occurs for the current at relatively low frequency ($f < 100$ Hz) where the Ohmic component is predominant. At higher frequency, the capacitance component prevails. In this case, the increase of the current is due to reduction of the resistance of the nematic system (decrease of both real and imaginary part of the complex impedance, as we have further verified by means of impedance spectroscopy).

Figure 3(c) compares the *I-V* behaviors obtained at a low-frequency (8 Hz) AC field (in order to exploit the pure Ohmic effect) for the same EPH-doped $Aerosil_{300}$/7CB nanogel in two cells of identical thickness, but either with PI or without PI layers (ITO only), both measured either without and with UV light illumination, under identical other experimental conditions. As seen, the examined films exhibit linear *I-V* dependence (Ohmic conductance). It should be noted that at 8 Hz the linear *I-V* dependence holds up to 100 V_{RMS}. Under the present level of the UV light intensity, there is a significant difference for the UV-induced increase factor of the AC conductivity ($\Delta\sigma_{UV}$) measured from the *I-V* behaviors of both samples. An important result is that $\Delta\sigma_{UV}$ is larger (~ 2) for EPH-doped $Aerosil_{300}$/7CB films with PI layers than that measured for the same nanocomposite films with ITO glasses only (figure 3 c, pairs (i) and (ii), respectively). Indeed, in the latter case $\Delta\sigma_{UV}$ was negligible (1.04 – 1.05) - a detectable change but hardly discernible within the experiment uncertainty. This significant difference is strongly indicative of the generation of free carriers in our EPH-doped $Aerosil_{300}$/7CB nanogel through the impinged PI layer after absorption of the UV light energy.

4. Conclusion

For EPH-photo-sensitized $Aerosil_{300}$/7CB nanocomposite films studied here, the surface orienting effect of unidirectional rubbed PI alignment layers is of paramount importance for the UV photo-driven stimulation of their electro-optics. The large UV light-induced change in the transparency of 3 wt.% nano(aerosil)-filled nematic doped with 3 wt.% azobenzene photoactive molecules (in our case EPH, a mesogen itself) is closely related to the alignment owing to the rubbed PI layers. In our $Aerosil_{300}$/7CB nanocomposite films in planarly-orienting cells, the *cis*-isomers formed by UV irradiation modify the LC anchoring conditions. An UV light-induced change of anchoring conditions due to space charge generation by the PI layers (close to the PI surfaces) is also possible to contribute to the UV-light-modified electro-optics of such photoresponsive nematic systems, but the role of this mechanism is much less.

Acknowledgments
The authors are highly indebted to Academician Prof. DSci. Alexander G. Petrov for his keen interest in this work and also for many useful discussions and support. The authors acknowledge the INERA EU project FP7 REGPOT-2012-2013-1 for their participation at INERA Workshop "Membrane and Liquid Crystals Nanostructures", 3-6 September, 2016, Varna, Bulgaria. Work supported by research project DFNI-TO2/18 by the Ministry of Education and Science, National Science Fund of Bulgaria.

References

[1] Iannacchione G S 2004 *Fluid Phase Equlibria* **222-223** 177
[2] Kreuzer M and Eidenschink R 1996 Filled Nematics *Liquid Crystals in Complex Geometries*

Formed by Polymer and Porous Networks eds G P Crawford and S Zumer (London: Taylor & Francis) chapter 15 p 307

[3] Bellini T, Buscaglia M, Chiccoli C, Mantegazza F, Pasini P and Zannoni C 2000 *Phys. Rev. Lett.* **85** 1008

[4] Jin T and Finotello D 2001 *Phys. Rev. Lett.* **86** 818

[5] Marinelli M, Ghosh A K and Mercuri F 2001 *Phys. Rev.* E **63** 061713

[6] Kutnjak Z, Kralj S and Zumer S 2002 *Phys. Rev.* E **66** 041702

[7] Roshi A, Iannacchione G S, Clegg P S and Birgeneau R J 2004 *Phys. Rev.* E **69** 031703

[8] Lobo C V, Prasad S K and Yelamaggad C V 2006 *J. Phys. Condens. Matter* **18** 767

[9] Marinov Y G, Hadjichristov G B, Petrov A G and Prasad S K 2016 *Composites B Engineering* **90** 471

[10] Marinov Y G, Hadjichristov G B, Petrov A G and Prasad S K 2016 *J. Phys. Conf. Ser.* **682** 012030

[11] Ciuchi F, Giocondo M and Barberi R 2008 *Liq. Cryst.* **35** 99

[12] O'Neill M and Kelly S M 2000 *J. Phys. D Appl. Phys.* **33** R67

[13] Hwang S J, Chen T A, Lin K R and Jeng S C 2012 *Appl. Phys. B* **107** 151

[14] Sridevi S, Hiremath U S, Yelamaggad C V, Prasad S K, Marinov Y G, Hadjichristov G B and Petrov AG 2011 *Mater. Chem. Phys.* **130** 1329

[15] Hadjichristov G B, Marinov Y G and Yelamaggad C V 2014 *J. Phys. Conf. Ser.* **558** 012026

[16] Kumar M V, Prasad S K, Marinov Y, Todorova L and Petrov A G 2015 *Mol. Cryst. Liq. Cryst.* **610** 51

[17] Prasad S K, Nair G G, Hegde G, Sandhya K L, Rao D S S, Lobo C V and Yelamaggad C V 2005 *Phase Transitions* **78** 443

[18] Petrov A G, Marinov Y G, Hadjichristov G B, Sridevi S, Hiremath U S, Yelamaggad C V and Prasad S K 2011 *Mol. Cryst. Liq. Cryst.* **544** 3[991]

[19] Yelamaggad C V, Prasad S K and Li Q 2012 Photo-Stimulated Phase Transformations in Liquid Crystals and Their Non-Display Applications *Liquid Crystals Bbeyond Displays: Chemistry, Physics, and Applications* ed Q Li (Hoboken, NJ: Wiley & Sons) chapter 5 p 157

[20] Natansohn A and Rochon P 2002 *Chem. Rev.* **102** 4139

[21] Yager K G and Barrett C J 2006 *J. Photochem. Photobiol. A* **182** 250

[22] Yager K G and Barrett C J 2009 Azobenzene Chromophores in Photo-Reversible Materials *Smart Light-Responsive Materials: Azobenzene-Containing Polymers and Liquid Crystals* eds Y Zhao and T Ikeda (Boca Raton, Florida: John Wiley and Sons) chapter 1 p 1

[23] Bellini T and Clark N A 1996 Light scattering as a probe of Liquid crystal Ordering in silica aerogels *Liquid Crystals in Complex Geometries Formed by Polymer and Porous Networks* eds G P Crawford and S Zumer (London: Taylor & Francis) chapter 19 p 381

[24] Mercuri F, Ghosh A K and Marinelli M 1999 *Phys. Rev. E* **60** R6309

[25] Caggioni M, Roshi A, Barjami S, Mantegazza F, Iannacchione GS and Bellini T 2004 *Phys. Rev. Lett.* **93** 127801

[26] Lu J, Deshpande S V, Gulari E, Kanicki J and Warren W L 1996 *J. Appl. Phys.* **80** 5028

[27] Jeng S C and Hwang S J 2012, Controlling the Alignment of Polyimide for Liquid Crystal Devices *High Performance Polymers - Polyimides Based - From Chemistry to Applications* ed M J M Abadie (Rijeka, Croatia: InTech,) chapter 5 p 87

[28] Kim J H, Kumar S and Lee S D 1998 *Phys. Rev. E* **57** 5644

[29] Scaramuzza N, Strangi G and Versace C 2001 *Liq. Cryst.* **28** 307

[30] Andrienko D, Kurioz Y, Nishikawa M, Reznikov Y and West J L 2000 *Jpn. J. Appl. Phys.* **39** 1217

[31] AkiyamaH and Iimura Y 2002 *Jpn. J. Appl. Phys.* **41**, Part 2, L521

[32] Sasaki T, Tsumura M, Nagae Y, Suzuki M and Iwata T 1995 *Electr. Commun. Jpn.* **78**, 79

[33] Matsuda H, Seo D S, Yoshida N, Fujibayashi K, Kobayashi S and Yabe Y 1995 *Mol. Cryst. Liq. Cryst.* **264** 23

IOP Conf. Series: Journal of Physics: Conf. Series **780** (2017) 012009 doi:10.1088/1742-6596/780/1/012009

Dielectric study of azo-doped aerosil/7CB nematic nanocomposite upon UV light

Y G Marinov [1], M P Marinov [1], G B Hadjichristov [1,*], A G Petrov [1], S K Prasad [2], L Marino [3] and N Scaramuzza [3]

[1] Georgi Nadjakov Institute of Solid State Physics, Bulgarian Academy of Sciences, 72 Tzarigradsko chaussee, Sofia 1784, Bulgaria

[2] Centre for Nano and Soft Matter Sciences, P.O.Box 1329, Jalahalli, Bangalore 560 013, India

[3] Dipartimento di Fisica, Università della Calabria, Via Pietro Bucci, Cubo 31C, 87036 Rende (CS), Italy

E-mail: georgibh@issp.bas.bg

Abstract. We studied three-component nanocomposite produced from nano-filled nematic by doping with photoactive azobenzene liquid crystal 4-(4'-ethoxyphenylazo)phenyl hexanoate (EPH) at concentration of 3 wt.%. The photo-insensitive host nanocomposite material (the filled nematic) was a gel formed from the nematic 4-n-heptyl cyanobiphenyl (7CB) and 3 wt.% of silica nanospheres (hydrophilic Aerosil 300) of size ca. 7 nm. The electrochemical impedance spectroscopy was applied to determine the UV-light-produced effect in EPH-doped aerosil/7CB upon continuous *trans*-to-*cis* photoisomerization of EPH molecules by varying the UV light intensity. The effect from UV illumination on the dielectric permittivity function of the studied photoresponsive nematic nanocomposite was analyzed at a room temperature in the frequency range 0.5 Hz – 200 kHz.

1. Introduction

Investigations of systems consisting of silica nanoparticles (NPs) in anisotropic liquid crystal (LC) media have become a large area of research primarily due to their electric and dielectric properties useful for prospective applications in electronic device industry [1-5]. A numerous dielectric studies on dispersions of ANPs in nematic LCs, in particular cyanobiphenyl-based compounds, have been reported [6-16]. Here, we present a dielectric study of photoresponsive nanostructured nematic material produced from a soft gel nanocomposite of nematic LC heptylcyanobiphenyl (7CB) and 3 wt.% hydrophilic aerosil NPs (ANPs) [16,17] by inclusion of photoactive azobenzene-containing LC 4-(4'-ethoxyphenylazo)phenyl hexanoate (EPH). Our aim is to inspect how the dielectric properties of EPH-doped aerosil/7CB thin layers (in particular with a thickness of 25 μm) are changed when the layers are exposed to illumination with continuous UV light source. In order to study the light-controllable change of the dielectric permittivity due to an efficient *trans*-to-*cis* photoisomerization in the studied photo-sensitized nematic nanomaterial, dielectric spectroscopy was applied in the frequency range 0.5 Hz – 200 kHz.

* To whom any correspondence should be addressed.

2. Experimental

ANPs (hydrophilic Aerosil 300 nanospheres having a mean diameter of about 7 nm and capped with silanol groups) were dispersed at a concentration of 3 wt.% in the LC 4'-heptyl-4-biphenylcarbonitrile (7CB) that exhibits a nematic phase at a room temperature. The ANPs/7CB nanocomposite [17] was doped with the azo-containing LC 4-(4'-ethoxy phenylazo)phenyl hexanoate (EPH) at a concentration of 3 wt.%. This compound was synthesized in the Centre for Nano and Soft Matter Sciences in Bangalore, India, and was specially chosen in order to obtain a high degree of UV light-induced isomerism. The maximum of the absorption of EPH is in the wavelength range from 360 nm to 380 nm. The nematic 7CB is optically transparent in this wavelength range. A sample of EPH-doped ANPs/7CB soft gel with a thickness of 25 μm was formed in a cell assembled from two ITO (indium tin oxide)-coated glass plates (to serve as electrodes) overcoated with a rubbed polyiamide orienting layers.

The EPH-doped ANPs/7CB sample was illuminated with UV continuous light source - a LED (λ = 375 nm). The UV light intensity on the sample was fixed at ~ 3 mW/cm^2 and kept the same by all the measurements in this work. The measurement of electrical impedance of the sample was performed by Potentiostat/Galvanostat SP-200 (product of Bio-Logic), set in a two-electrode configuration. The complex impedance as a function of the frequency (*f*) of the applied electric field was recorded in the frequency range from 0.5 Hz to 0.2 MHz. The amplitude of AC voltage applied on the cell was 2.5 V. During the measurements, identical experimental conditions were kept, only increasing the exposure dose of the continuous UV illumination. For the sake of rapid measurements, no averaging of the data was undertaken during the recording of the impedance spectra. The time interval between data acquisition was 0.6 s, the single run was within 85 s (140 experimental points in 5 decades of the frequency scan). The temperature of the sample was 23° C.

3. Results and Discussion

Figure 1 shows the raw data obtained for the complex impedance of EPH-doped aerosil/7CB film being exposed to UV light.

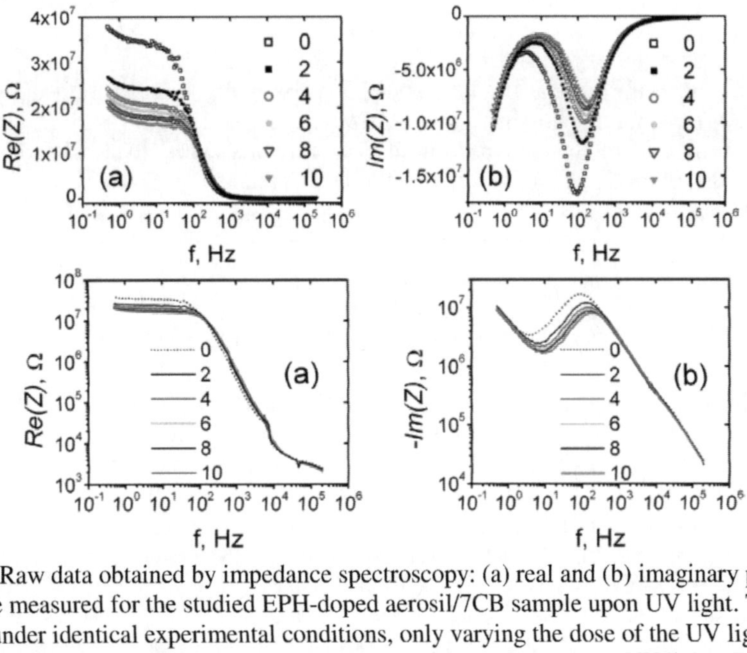

Figure 1. Raw data obtained by impedance spectroscopy: (a) real and (b) imaginary parts of complex impedance measured for the studied EPH-doped aerosil/7CB sample upon UV light. The spectra were recorded under identical experimental conditions, only varying the dose of the UV light exposure (the labels in the graph denote the time of exposure in minutes; 0 means no UV light). Down: the same, but in logarithmic scale.

The spectra were somewhat noised because of fast data acquisition (and no averaging of the data; respectively). This was necessary in order to quickly record and adequately monitor the change of the system in several distinguished steps during the *trans-cis* photoisomerization before the saturation. For the same reason, the intensity of the UV illumination was chosen to be relatively low. Even in this case, it can be seen from figure 1 that by slightly increasing dose of the UV illumination the response of the examined azo-doped nanofilled nematic is distinctly changed. The change is a little indeed, but well detectable. In the low frequency region (below ~ 100 Hz) both real (*ReZ*) and imaginary (*ImZ*) parts of impedance clearly decreases with the increasing UV light dose.

From impedance data one can obtain the dielectric permittivity, and thereby an information about the change of dielectric properties of the examined system upon UV light. The real (ε′) and imaginary (ε″) parts of the frequency-dependent complex dielectric function ε* = ε′ – i ε″ were calculated using the acquired ReZ and ImZ by the same procedure used in Refs. [18-20]. As an example, figure 2 reports dielectric spectra calculated for the initial state of the azo-doped system (no UV light) and by UV illumination at the highest dose we have used here, in order to be compared.

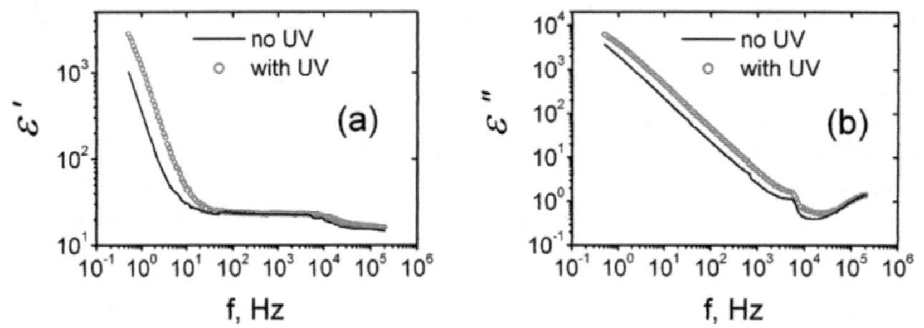

Figure 2. Real (a) and imaginary (b) parts of complex dielectric function calculated from data in figure 1 for the case with no UV light (lines) and with UV light at the highest dose (10 min exposure) used here (circles).

The analysis of dielectric spectra for the complex permittivity was performed by use of a combination of both Kohlrausch-Williams-Watts (KWW) model [21,22] and Havriliak-Negami (HN) model [23,24], with addition of conductivity contribution present at lower frequencies. This approach is an empirical modification and generalization of the Debye relaxation model. Thus, the following fitting function was used:

$$\varepsilon^* = \frac{\sigma}{\varepsilon_0 (2\pi f)^n} + F.T.\left\{ \Delta\varepsilon_{KWW} \exp\left[-\frac{t}{\tau} \right]^{\beta} \right\} + \frac{\Delta\varepsilon_{HN}}{\left[1 + i (f / f_0)^a \right]^b} \qquad (1)$$

with a term of conductivity (the first term), that of KWW model (the second term) and the product model HN (the third term). In Eq. (1), f is the frequency of the applied electric field, f_0 is the relaxation frequency, σ is the DC electrical conductivity, ε_0 is the vacuum permittivity, $\Delta\varepsilon$ is the dielectric strength, n is the power factor of the conductivity term. The parameters a and b account for the shape of the dielectric dispersion curve, *F.T.* means the Fourier transform, β is the stretching exponent, τ is the time for dielectric relaxation, and t is the time variable. For the nematic soft gel system under study, the inclusion of KWW function (proper for analysis of data for dielectric permittivity of polymers [25]) in the fitting function should be more appropriate, than the use of HN relaxation function only [26] (usually applied for modeling of dielectric spectra of LCs). Representative fits of the modulus (ε) of ε^* by use of the combined model are shown in figure 3.

Table 1 summarizes the values of parameters corresponding to the best fit of the permittivity. In the last columns are the values of the <reduced χ^2> coefficient and the regression coefficient ($\Re$) for the fitted lines. As seen, these two values, for each fit, are excellent. Our calculated results allows to do qualitative conclusions for the change of the azo-doped nanofilled nematic due to absorption of UV light energy, followed by the corresponding *trans*-to-*cis* photoisomerization of the azo dopants (the transformation from rod-like *trans* to bent-shaped *cis* conformers of the EPH molecules). In this way one can illustrate what happens in the azo-doped nanogel when the *cis*-to-*trans* ratio permanently increases.

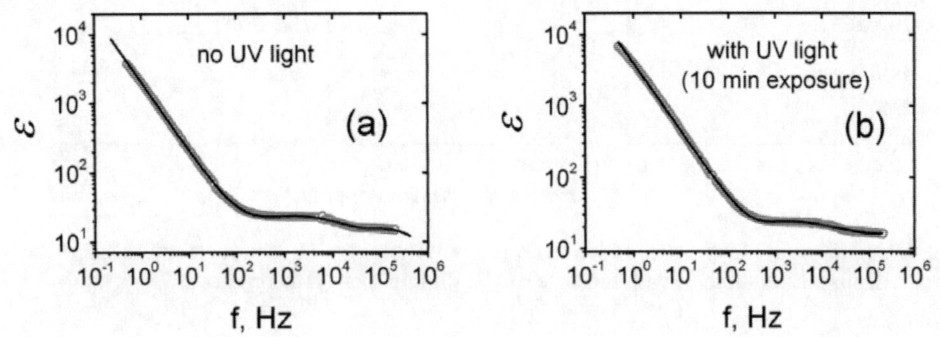

Figure 3. Dielectric spectra of ε calculated for the studied EPH-doped aerosil/7CB sample in the case with: (a) no UV light; (b) with UV light at the highest dose (10 min exposition) we have used. The lines represent the best fits with a combined complex function according to Eq. (1).

Table 1. Best-fit parameters by modelling with a combined complex function according to Eq. (1).

UV [min]	Log(σ)	n	KWW-log(τ)	KWW-Δε	KWW-β	HN-log(f_o)	HN-Δε	HN-a	HN-b	⟨χ^2⟩	$\Re$
0	-8.954	0.984	-6.211	22.062	0.759	3.414	100.347	0.518	1.088	0.0134	0.9998
2	-8.815	0.973	-4.451	100.354	0.378	4.865	561.937	1.345	0.005	0.0117	0.9999
4	-8.767	0.962	-4.445	101.488	0.400	4.892	83.222	1.371	0.038	0.0276	0.9997
6	-8.763	0.968	-5.459	120.289	0.250	6.498	26.301	0.487	42.931	0.0213	0.9998
8	-8.738	0.965	-5.504	120.672	0.250	5.641	26.644	0.503	17.401	0.0221	0.9998
10	-8.687	0.955	-8.755	158.789	0.278	2.876	125.354	0.820	0.231	0.0264	0.9997

At lower frequencies, the movement of charge carriers via the sample becomes important and this is reflected in the dielectric spectra. The simple movement of charge carriers (electrical conduction) leads typically to a response that is similar to the one of an Ohmic conductor. The values of electrical conductivity σ and the power factor *n* obtained by the fits of dielectric spectra are shown in table 1 and figure 4 (a,b). Although the results deviate from monotonic function (most probably caused from the noised spectra of the input data – the impedance initially measured), a clear increasing trend takes place for σ (figure 4 a). This is reasonable in view of the reduction of *ReZ* with the UV dose, well seen in the low frequency region (recall figure 1 a). Most likely the change of σ can be even approximated with a linear dependence as a function of the increasing UV dose. A possible explanation of the resistivity reduction of EPH-doped aerosil/7CB sample upon UV light could be that the transformation from rod-like *trans* to bent-shaped *cis* conformers of the EPH molecules introduce some disorder in the system. Such a state corresponds to a state that can be characterized with a somewhat higher temperature. It is known for the ionic electrolytes [27] that the higher temperature, the higher conductivity.

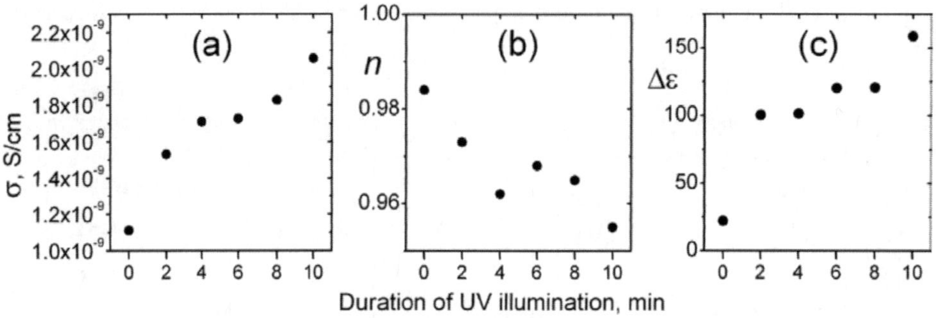

Figure 4. Variations of electrical conductivity σ (a), power factor *n* (b) and dielectric strength Δε (c) (from data in table 1) with the dose of UV illumination (0 means no UV light).

The power factor *n* calculated by the fits of the dielectric spectra exhibits a decreasing behaviour with the increasing dose of UV illumination (figure 4 b). Thus, some qualitative conclusions can be done, for example about the change of the homogeneity of the system (note, such a change is only partially reflected in the value of the parameter *n*). As known [28], the reduction of *n* can be interpreted as a deviation from the single (fundamental) dielectric mode at a defined single frequency. By that, the fundamental mode for which $n = 1$ (ideal dielectric material) is spread into a lot of characteristic frequencies corresponding to a variety of dielectric modes. This spread could be associated with the increase of the inhomogeneity (in particular due to disordering) of the dielectric material.

The *cis*-isomers of EPH molecules formed by UV irradiation, being not-nematogenic, eventually introduce some decrease of the degree of nematic order in EPH-doped ANPs/7CB gel similar to other LCs containing azobenzene units when exposed to UV radiation, e.g., [29]. Such disordering effect should be relatively weak due to the relatively low concentration of the azo-molecules (3 wt % in our case). Contrary to the inclusion of azo-molecules in pure nematic liquid crystal, the effect from the photo-induced disorder after the *trans-cis* isomerization of the azo dopants should be relatively weaker in the gel under study, being actually a strongly disordered system itself at the present concentration (3 wt.%) of ANPs forming chains in a soft-gel nematic material [16,17].

The disordering can not be directly evidenced with dielectric spectroscopy, but such effect may result in a shift of the principal dielectric peak of 7CB at higher frequency (2 – 6 MHz) [30]. Unfortunately, within the maximum frequency limit of the apparatus employed here one can only partially register and see this dielectric mode in our spectra. The same peak at higher frequencies (in the range $10^6 - 10^7$ Hz) is also present in pure 7CB (without silica nanoparticles) and this represents a molecular process associated with the rotation of the 7CB molecules around their short molecular axis [31]. The doping with silica nanoparticles introduces some effects that influence the same process [8].

More important would be the evidence of the UV light promoted change in the dipolar character of the system, a property that should be strongly affected by the UV illumination (much stronger than the eventual UV-light-induced disorder of the gel through the EPH molecules that in our case decorate the gel network). It is well known that the *cis*-isomers of azobenzene compounds possess an electrical polarity (in contrast to their *trans*-isomers). Thus, if the UV light intensity is high enough, the dipole moment of the azobenzene group can be switched between approximately from 0 (for *trans*) to 3 D (for *cis* conformers). Actually, since the conductivity change is involved, the fitting parameter n accounts for the contributions of both phenomena (dipolar and disordering), but other factors play also a role. The parameter n has a value of 1 for a pure Ohmic conduction. Also $n = 1$ for an ideal dielectric material. In the context of the dipolar property of the material, the reduction of n (recall figure 4 b) can be interpreted as a deviation from the ideal dielectric due to creation of UV-light-induced dipoles within the material volume (or due to influence of electrode surface polarization).

A more reliable indicator of the enhanced dipolar character of the EPH-doped aerosil/7CB nanofilled nematic gel upon UV light is the increase of the dielectric strength $\Delta\varepsilon$. This fit parameter is directly related to the dipole changes, that in our case do result from the *trans*-to-*cis* photoisomerization of the azo-bonded EPH dopants. Figure 4 (c) reports the variation of $\Delta\varepsilon$ with the UV dose. The dielectric increment can be attributed to the UV-light-induced increasing of the dipole moment of each EPH molecule and hence it implies the enhanced electro-dipolar character of the system upon UV light. Also, the same applies to the increase of both ε' and ε'' (and thus the modulus of ε^*) with the UV dose (figure 2). Our analysis indicated that a nearly monotonic increase with the UV dose takes place, that is reasonable when the material becomes more dipolar (the dielectric constants increase with molecular dipole moment and polarizability, e.g. [32,33].

4. Conclusion

Using data collected by impedance spectroscopy, we have experimentally characterized the change of dielectric properties of azo-doped nanofilled nematic (in particular, EPH-doped aerosil/7CB gel) upon continuous UV light. The information deduced by dielectric spectroscopy analysis was related to the macroscopic properties of the examined photoresponsive nanocomposite. Note that the UV-light-induced change of the *cis*-to-*trans* ratio can be easily evidenced by optical spectroscopy (through the measurement of the alteration of absorption bands relevant to both *trans* and *cis* conformers) (also in real time). However, the additional information obtained by dielectric spectroscopy about the change of the dipolar character of the system (and for the conductivity in the same time) during the UV illumination, should be also of importance. The dielectric spectroscopy could be sensitive for such photo-induced changes and can be used to achieve reliable conclusions (in our case the conclusions are more or less qualitative because of integral character of the dielectric measurements). In particular, the situation in is rather complicated due to the fact that various processes are involved and many factors take place by photo-induced changes in the studied azo-doped nanofilled nematic, that affect the system parameters and the material properties of this photoresponsive material.

In our experiment, the use of single-pulsed UV-light radiating lasers with nanosecond pulse duration (in particular, by third-harmonic radiation of neodymium lasers at 355 nm that also induces *trans*-to-*cis* photoisomerization of EPH molecules) enables more precise control of the portion of UV radiation imposed on the sample. Thus, the considered *trans-cis* photo-induced effect by the studied azo-doped nanocomposite gel can be finely tailored, and thereby the behaviors of fitting parameters σ, n and $\Delta\varepsilon$ as a function of the UV-light dose could be more accurately obtained (work in progress).

Acknowledgments
The authors acknowledge the INERA EU project Research Potential "Research and Innovation Capacity Strengthening of ISSP-BAS in Multifunctional Nanostructures" (FP7 REGPOT-2012-2013-1) for their participation at the INERA Workshop “Membrane and Liquid Crystal Nanostructures” (MELINA 2016), 3-6 September, 2016, Varna, Bulgaria. Work supported by research project DFNI-TO2/10 by the Ministry of Education and Science, National Science Fund of Bulgaria.

References

[1] Dolgov L and Yaroshchuk O 2004 *Colloid Polym. Sci.* **282** 1403
[2] Garbovskiy Y A and Glushchenko A V 2011 Liquid Crystalline Colloids of Nanoparticles: Preparation, Properties and Applications *Solid State Physics*, vol **62**, eds R E Camley and R

L Stamps (New York: Academic Press, Elsevier) p 1

[3] Chaudhary A, Malik P, Mehra R and Raina K K 2012 *Phase Transitions* **85** 244

[4] Li Q 2012 *Liquid Crystals Beyond Displays: Chemistry, Physics, and Applications* (Hoboken, NJ: Wiley & Sons)

[5] Blanc C, Coursault D and Lacaze E 2013 *Liq. Cryst. Rev.* **1** 83

[6] Tschirske S, Yaroshchuk O V and Kresse H 1995 *Cryst. Res. Technol.* **30** 571

[7] Yaroshchuk O, Kovalchuk O and Zakrevska S 2000 *Funct. Mater.* **7** 732

[8] Hourri A, Bose T K and Thoen J 2001 *Phys. Rev. E* **63** 051702

[9] Frunza S, Frunza L, Goering H, Sturm H and Schönhals A 2001 *Europhys. Lett.* **56** 801

[10] Sinha G, Leys J, Glorieux C and Thoen J 2005 *J. Non-Cryst. Solids* **351** 2780

[11] Sinha G, Leys J, Glorieux C and Thoen J 2005 *Phys. Rev. E Stat. Nonlin. Soft. Matter Phys.* **72** 051710

[12] Lobo C V, Prasad S K, Yelamaggad C V 2006 *J. Phys.: Condens. Matter* **18** 767

[13] Różański S A , Sinha G P and Thoen J 2006 *Liq. Cryst.* **33** 833

[14] Jayalakshmi V, Nair G G and Prasad S K 2007 *J Phys Condens Matter* **19** 226213

[15] Yelamaggad C V, Prasad S K and Li Q 2012 Photo-Stimulated Phase Transformations in Liquid Crystals and Their Non-Display Applications *Liquid Crystals Beyond Displays: Chemistry, Physics, and Applications* ed Q Li (Hoboken, NJ: Wiley & Sons) chapter 5 p 157

[16] Kumar M V, Prasad S K, Marinov Y, Todorova L and Petrov A G 2015 *Mol. Cryst. Liq. Cryst.* **610** 51

[17] Marinov Y G, Hadjichristov G B, Petrov A G and Prasad S K 2016 *Composites B: Engineering* **90** 471

[18] Marino L, Ionescu A T, Marino S and Scaramuzza N 2012 *J. Appl. Phys.* **112** 114113

[19] Marino L, Ionescu A T, Pucci D, Marino S, Bruno E and Scaramuzza N 2014 *J. Appl. Phys.* **116** 114101

[20] Carbone F, Mazzulla A, Ciuchi F and Scaramuzza N 2015 *Europ. Phys. J. Plus* **130** 151

[21] Kohlrausch R 1854 *Ann. Phys. Chem. (Poggendorff)* **91** 179

[22] Williams G and Watts D C 1970 *Transactions of the Faraday Society* **66** 80

[23] Havriliak S and Negami S 1966 *J. Polym. Sci. C* **16** 99

[24] Havriliak S and Negami S 1967 *Polymer* **8** 161

[25] Alzari V, Nuvoli D, Sanna V, Caruso T, Marino S and Scaramuzza N 2016 *AIP Advances* **6** 035005

[26] Havriliak S, Jr. and Havriliak S 1994 *J. Non-Cryst. Solids* **172-174** 297

[27] Bockris J O'M, and Reddy A K N 1998 *Modern Electrochemistry: Ionics* vol **1** 2nd ed (Nework: Plenum Press)

[28] Almond D P and West A R 1983 *Solid State Ionics* **9-10** 277

[29] Dubtsov A, Pasechnik S, Shmeliova D and Kralj S 2014 *Appl. Phys. Lett.* **105** 151606

[30] Haase W and Bormuth F J 1989 The Significance of Dielectric Relaxation Processes and Switching Properties for Data Storage in Liquid Crystal Polymer Media *Polymers in Information Storage Technology* ed K L Mittal (New York: IBM US Technical Education) p 51

[31] Bose T K, Campbell B, Yagihara S and Thoen J 1987 *Phys. Rev. A* **36** 5767

[32] Smyth C P 1955 *Dielectric Behavior and Structure: Dielectric Constant and Loss, Dipole Moment and Molecular Structure* (New York: McGraw-Hill)

[33] Scaife B K P 1998 *Principles of Dielectrics (Monographs on the Physics & Chemistry of Materials)* 2 nd ed (Oxford: Oxford University Press)

IOP Conf. Series: Journal of Physics: Conf. Series 780 (2017) 012010 doi:10.1088/1742-6596/780/1/0

Surface characterization and orientation interaction between diamond- like carbon layer structure and dimeric liquid crystals

H Naradikian[1], M Petrov[1], B Katranchev[1], T Milenov[2] and S Tinchev[2]

[1] Institute of Solid State Physics, Bulgarian Academy of Sciences, 72 Tzarigradsko chaussee Blvd., 1784 Sofia, Bulgaria.
[2] Institute of Electronics, Bulgarian Academy of Science, 72 Tzarigradsko chaussee Blvd., 1784 Sofia, Bulgaria.

E-mail: harry@issp.bas.bg

Abstract. Diamond-like carbon (DLC) and amorphous carbon films are very promising type of semiconductor materials. Depending on the hybridization sp^2/sp^3 ratio, the material's band gap varies between 0.8 and 3 eV. Moreover carbon films possess different interesting for practice properties: comparable to the Silicon, Diamond like structure has 22-time better thermal conductivity etc. Here we present one type of implementation of such type nanostructure. That is one attempt for orientation of dimeric LC by using of pre-deposited DLC layer with different ratio of sp^2/sp^3 hybridized carbon content. It could be expected a pronounced π_1-$\pi_2$interaction between *s* and *p* orbital levels on the surface and the dimeric ring of LC. We present comparison of surface anchoring strengths of both orientation inter-surfaces DLC/dimeric LC and single wall carbon nanotubes (SWCNT)/dimeric LC. The mechanism of interaction of dimeric LC and activated surfaces with DLC or SWCNT will be discussed. In both cases we have π-π interaction, which in combination with hydrogen bonding, typical for the dimeric LCs, influence the LC alignment. The Raman spectroscopy data evidenced the presence of charge transfer between contacting hexagonal rings of DLC and the C = O groups of the LC molecules.

1. Introduction

The alignment process of liquid crystal (LCs) is of significant importance for fundamentalresearchers aiming to understand the variety of phenomena observedat the interface liquid crystal (LC) – aligning surfaces. Generally, thenematic LC (NLC) director $\boldsymbol{n}$ (the average direction of the longer axisof the LC molecules) can be collinear with the LC cell boundary surfacenormal (z axis) or perpendicular to it, thus determining the two typical surfacesymmetries, homeotropic and planar respectively [1].

In the case $\boldsymbol{n}\perp z$, $\boldsymbol{n}$ follows all possible directions overthe boundary *xy* surface plane.However, because of the anisotropic surfaceproperties, practically an "easy" axis appears. It defines a specialdirection ($\boldsymbol{n}_o$), determined by an unit vector at the interface plane,which minimizes the interface free energy.The "easy" direction can be created inmanyways. It exhibits the "anchoring" of $\boldsymbol{n}_o$ on the interface. Themost popularone is the rubbing of the polymer (e.g. polyimide) surface with specialcloths or SiOx oblique evaporation, thus creating an orienting surfacewhich causes$\boldsymbol{n}_o$ to align along the rubbing direction. Recently we applied a new method for surface orientation of LCs, by imposing unidirectional oriented single wall carbon nanotubes (SWCNTs) and found that such deposited carbon nanostructures significantly increase both the strength of the surface "anchoring" and the surface memory effects [2, 3]. As an important result we indicated that the carbon structures, mainly due to the electronic π–π interaction, induce in the vicinity of the orienting surface a double electric layer (selective electric ion adsorption), which strongly influenced the surface forces. Thus, the surfaceorientation of the LC system can be modified using the carbon structures, such as,

SWCNTs,amorphous carbon (a-C) or diamond –like carbons (DLC), which provoke a selectiveion adsorption and trapping of LC bulk charges,thus leading to surface charge density increase,

The goal of the present work is to estimate efficiency of the a-C and DLC layers deposition on the alignment of the nematic (N) dimeric liquid crystals. The π–π impact ability of a-C andDLC material on formation of an N LC monocrystals will be indicated. Using the liquid crystal anchoring strength estimated at orienting substrates treated with a-C andDLC/ITO we will present a comparisonwith that of SWCNT/ ITOtreated surfaces [3]. Weanalyze the influence of the dimeric LC structure on the growthof oriented N domains at different a-C andDLC coatings.

Diamond-like carbon (DLC) and amorphous carbon films are very promising type of semiconductor materials. Depending on the hybridization sp^2/sp^3 ratio, the material's band gap varies between 0.8 and 3 eV. The primary desirable qualities of DLC are hardness, wear resistance, and slickness [4]. DLC is usually applied as coatings to other materials that could adopt some of those properties. The more common carbon atoms in DLC are in sp^3 hybridized state predominantly and they possess very high hardness and electrical resistivity. The hardness as well as the resistivity decreases with increasing of the sp^2 contentof the carbon film [5]. By mixing these types in various ways at the nanoscale level of structure, DLC and a-Cfilms can be made that at the same time are amorphous, flexible, thus presenting interesting material characteristics. The hardest, strongest, and slickest is such a mixture, known as tetrahedral amorphous carbon (ta-C). Such type of carbon structure can be considered to be the "pure" form of DLC, since it consists only of sp^3 bonded carbon atoms. The amorphous graphite films consists of predominantly sp^2 hybridized carbon without long-range ordering.

2. Experimental

The experiments have been carried out with*4-n*-alkoxybenzoic acids, nOBAs, which are widely used as components in supramolecularhydrogen bonded LCs. Twoof them with alkyl chain lengths of $n = 7$ and 8 have been exploited, which have the following phase transitions:

7OBA: Cr↔92°C SC↔98°C N↔146°C I;

8OBA: Cr↔101°C SC↔108 °C N↔147°C I;

Its chemical structure in dimeric form is demonstrated below:

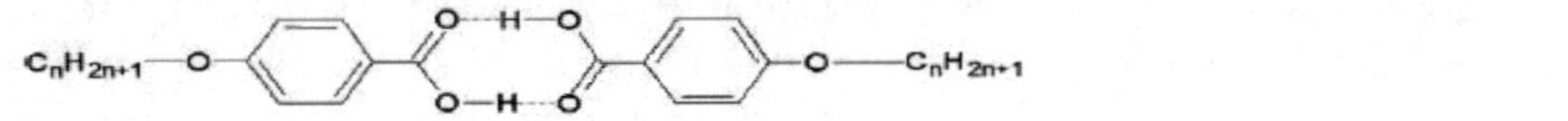

The LC material in its isotropic phase was filled in the cell using thecapillary force action. The gap between the glass plates, defining a cell thickness of 12 μm, was maintained with Mylar spacers. The temperatureof the sample was varied with a rate of 0.2 °C min^{-1} by a hotstage temperature controller Linkam TMS 90, and the chosen experimentaltemperature was stabilized with an accuracy of ±0.1 °C.Monitoringthe optical textures, in the liquid crystal cell (LCC)*xy* plane was carried out using a video camera Hitachi on a microscopeZeiss NU2.In turn, the optical signal was studied by video capturingequipment.

We deposited carbon thin films applying the plasma enhanced chemical vapor deposition (PECVD) method, which is widely used for coating of thin carbon films due to its high reliability. By this method the sp^2/sp^3 hybridized carbon ratio in the films can be varied with the plasma voltage. We deposited thin films on indium-thin oxide (ITO) coated glass substrates aiming of obtaining layers with different phase composition and thickness. It was also expected that the increased number of π-bonds in sp^2- dominated carbon films will enhance additionally the orientation of liquid crystals inserted on the carbon films.Several specimens were also nitrogen doped during the deposition in order to decrease the resistance of the carbon layer.

For the experiments we used four types of carbon films designated as follows: C_02, TCH_03, TC_04 and TCH_05 (the specimens TCH_03 and TCH_05 are doped with about 0.2-0.4 at% nitrogen). The thickness and phase composition of carbon films were determined by ellipsometric and X-ray photoelectron spectroscopy measurements carried on control samples. The XPS study enables distinguishing of sp^2 and sp^3 fractions [6, 7].The ellipsometry measurements were performed using J.A. Woollam Co., Inc. M2000D rotating compensator spectroscopic ellipsometer with a CCD spectrometer with wavelength range from 193 to 1000 nm. Experimental data for Ψ and Δ were acquired at angles of incidence of 55, 60, 65, 70 and 75 degrees and were modelled using the CompleteEASE Woollam Co., Inc. software. The thickness of C_02, TCH_03, TC_04 and TCH_05 specimens according to the ellipsometry studies is 12-14, 140, 70 and 20 nm, respectively. The X-ray photoelectron spectra were obtained using non-monochromatized Al K_{α} (1486.6 eV) radiation in a VG ESCALAB MK II electron spectrometer under base pressure of $1x10^{-8}$ Pa. The spectrometer resolution was calculated from the $Ag3d_{5/2}$ line with the analyzer transmission energy of 20 eV. The full width at half maximum (FWHM) of this line is 1eV. The spectrometer was calibrated against the $Au4f_{7/2}$ line (84.0 eV) and the samples' charging was estimated from C1s (285 eV) spectra from natural hydrocarbon contaminations on the surface. The accuracy of the binding energy (BE) measured was 0.2 eV. The photoelectron spectra of C1s, O1s and N1s lines of carbon films deposited on different substrates were recorded and corrected by subtracting a Shirley-type background and quantified using the peak area and Scofield's photoionization cross-sections. The sp^2/sp^3 ratio in different specimens is about 90% in C_02 and about 60% in TCH_03, TC_04 and TCH_05 specimens according to XPS results. The results of XPS and ellipsometry measurements are summarized in Table 1 and the carbon films can be classified as amorphous carbon (a-C): C_02 specimen as well as tetrahedral carbon (ta-C): specimen TC_04 and hydrogenated ta-C (ta-C:H): specimens TCH_03, TCH_05. The Raman spectroscopy and especially the dispersion and the intensity ratio of D and G bands are established to enable the determination of sp^2/sp^3 ratio in thefilms [8]. Our Raman measurements were carried out in the micro-Raman spectrometer HORIBA Jobin Yvon Labram HR 800 Visible with a He-Ne (633 nm) laser. The laser beam with 0.5 mW power wasfocused on a spot ofabout 1 μm in diameter on the studied surfaces, the spectral resolution being 0.5 cm^{-1} or better.The Raman spectroscopy measurements were performed directly on the thin films deposited on ITO/glass substrates. The observed features in figure 1 strictly correlate with the results of XPS and confirm the phase composition of the specimens according to [8,9].

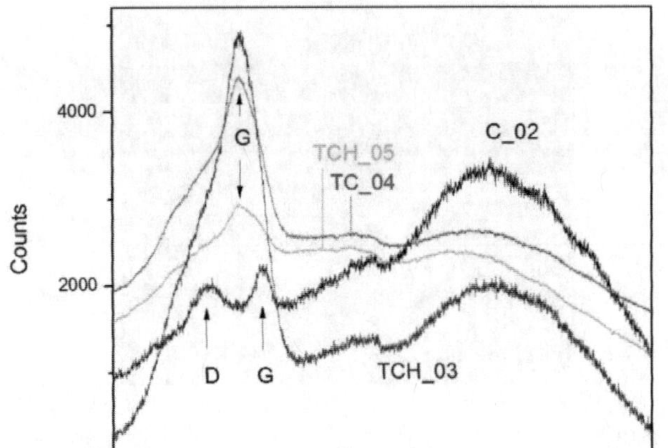

Figure 1. Raman spectra of C_02, TCH_03, TC_04 and TCH_05 specimens.The Raman spectrum of TCH_03, TC_04 and TCH_05 specimens is dominated by the G-band (that appears at about 1490-1500 cm^{-1}- (green, red and blue traces) while D and G bands (at 1332 and 1580 cm^{-1}, respectively) appear in the Raman spectrum of C_02 specimen- the black trace.

IOP Conf. Series: Journal of Physics: Conf. Series 780 (2017) 012010 doi:10.1088/1742-6596/780/1/012

Table 1. Summarized results for the layers configuration and thickness as well as for their phase composition obtained by XPS and ellipsometry studies.

Experiment	Fraction sp^2/sp^3 C	C, at%	O, at%	N, at%	Layer Thickness, nm
					ta-C:H
C_02	0.90	95.56	2.01	0.00	12-14
TC_04	0.55	92.19	4.48	0.33	70
TCH_03	0.60	92.65	5.03	0.23	140
TCH_05	0.62	93.01	4.01	0.40	20

3. Results and discussion

Choosing these specimens and by micro texture polarization analyze, we obtained the corresponding micro textures necessary for the anchoring energy calculation. In figure 2, the texture of C_02 specimen with 7OBA is indicated. We present the polarization analysis of the π-inversionwalls obtained by cooling from the isotropic phase at a cooling rateof 1 °C min^{-1}, used for the measurement of the wall's width *l*. The walls in crossed polarizers (P⊥A), where P denotes the polarizer and A the analyzer are seen. As is seen in figure 2, the π inversion walls,due to their meta-stability, can form irregularly shaped loops. Depending on the cooling rate elliptical wall loop also can be formed.Following the microtexture polarization analysis one observes in the figure the walls as bright lines.Basing on the polarization micro- textural analysis we assumethe observed wall to be Neel type. Hence we can usethe formula [1,2]$W_s=\pi^2Kd/2l^2$, where *d* is the cell thicknessand *K* is the elastic constant (10^{-11} J m^{-1})considered in theone-constant approximation. Here we assume *K* to represent thetrace of the elasticity tensor K_{ii} which is relevant for the Neel typewall, i.e. accounts for the director's deviations along polar and azimuthalcoordinates, leading to splay and bend shape of the wall.The wall thickness *l* was derived from a study of normally incident monochromatic light on

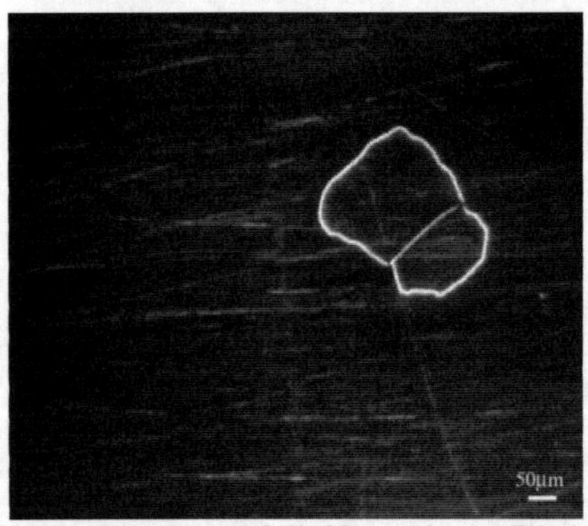

Figure 2. The micro texture of C_02 specimen filled with 7OBA. The π-inversion walls at T=150 °C in the N phase obtained at cooling rate 1 °C min^{-1} and cell thickness *d*=12 μm for P⊥A.

micrometric etalon.Using the measured*l* values for C_02 specimen and the Rapini-Papoular formula [2] $W_s=K(\boldsymbol{n}\cdot\boldsymbol{n}_o)^2L^{-1}$, we determine the extrapolation length *L* which is ameasure for the anchoring strength [1]. Assuming that ($\boldsymbol{n}\cdot\boldsymbol{n}_o$)≈1,we estimated the anchoring energy for the basic surface treatmentsand found the extrapolation length of the specimen to be 1.68 μm. The corresponding extrapolation lengths for the specimens TCH_03/8OBA, TC_04/8OBAare 6.75μmand for TCH_05/7OBA is 4.3μm. The value of the anchoring energy for C_02 /7OBA is W_s=0.59×10−5 J m^{-2}, while for the TCH_03/8OBA, TC_04/8OBAit is 0.148 ×10−5 J m^{-}2, and for TCH_05/7OBA is

0.23×10−5 J m^{-2}. As an example of the lower anchoring we present that of TCH_03/8OBA, TC_04/8OBA indicated in figure 3. As is seen on the Table 2, the surface anchoring is biggest for C_02 /7OBA and lowest for both TCH_03/8OBA, TC_04/8OBA specimens. At the same time, interesting is that the surface anchoring of TCH_05/7OBA specimen range between those of the C_02 /7OBA and TCH_03/8OBA, TC_04/8OBA.

Table 2. Represent the calculated values of anchoring energy and extrapolation length for different type of orientation interfaces.

Orienting surface/LC	**Anchoring energy - W_s Jm^{-2}**	**Extrapolation Length-L μm**
C_02 /7OBA	0.59×10^{-5}	1.68
TCH_03/8OBA, TC_04/8OBA	0.148×10^{-5}	6.75
TCH_05/7OBA	0.23×10^{-5}	4.3
SWCNT/7OBA	$16\text{x}10^{-5}$	0.06

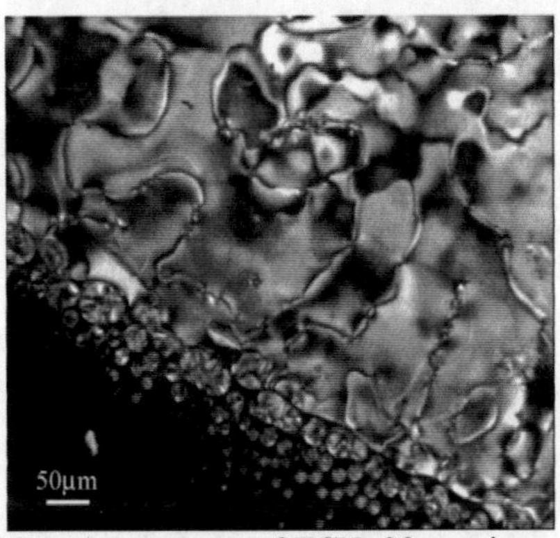

Figure 3. The micro texture of TCH_03 specimen filled with 8OBA. The π-inversion walls at T=169 °C in the N phase obtained at cooling rate 1 °C min^{-1} and cell thickness d=12 μm for P⊥A.

4. Conclusions

We test the favorable effect of themolecular structure on the N domain growth, and found that the p,n alkyloxybenzoic acids (nOBA) LCs, constituted by dimer rings, containing two parallel linear hydrogen bonds-cyclic dimer, are the most effective in this surface interaction. It is found that the dimeric LC/ta-C, LC/ta-C:H and LC/a-C anchoring energy is significantly less than that one of the dimeric LC/SWCNTs [3], and is approximately comparable with that of the conventional physicaladsorption, meaning thatthe LC/ta-C, LC/ta-C:H and LC/a-C coatings rather prefers the homeothropic than planar orientation. Thus, the main contribution to the surface energy anchoring strength is due to the combination of π-π/hydrogen bonds, both acting within the double electric surface layer, in direction perpendicular to the substrate's surface plane. The result also indicated that the decisive influence of the anchoring strength is that of the linear hydrogen bond energy which in 7OBA is rather bigger than that of 8OBA, i.e. 20kJ/mol and 5kJ/mol respectively [3]. We can generalized that the direct dimeric LC/ta-C, LC/ta-C: H and LC/a-C interaction can be controlled by π-π electronic bonding, strongly expressed in the sp^2 hybridization (the ratio sp^2 / sp^3), as the Raman spectra indicated. This investigation provokes the surface action of a-Cto be considered as a full picture and optimization of the influence of carbon nanostructures, SWCNT, DLC and a-Con the orientation of the LC.

Acknowledgments
The work is supported by grant INERA EU project (FP7-316309-REGPOT-2012-2013-1) "Research and Innovation Capacity Strengthening of ISSP-BAS in Multifunctional Nanostructures" and Grant BK-04-14 from the Institute of Solid State Physics, Bulgarian Academy of Sciences. The authors gratefully acknowledged Dr. Penka Terziiska (ISSP-BAS) for the ellipsometric measurements as well as Assoc. Prof. Dr. Ivalina Avramova (IGIC-BAS) for the XPS studies.

References

[1] de Gennes P G and Prost J 1993 *The Physics of Liquid Crystals* 2nd edition (New York: Oxford University Press)
[2] Rapini A, Papoular M 1969 *J. Phys. Colloq.* **30** 54
[3] Petrov M, Katranchev B, Rafailov P M, Naradikian H, Dettlaff-Weglikowska U and Keskinova E, 2013 *J. Mol. Liq.***180**, 215
[4] Robertson J 2002 *Mater. Sci. Engineer. R***37** 129
[5] Ferrari A and Robertson J 2004 *Phil. Trans. R. Soc. Lond.* A **362** 2477
[6] Lascovich J, Giorgi R and Scaglione S 1991 *Appl. Surf. Sci.* **47** 17
[7] Díaz J Paolicelli, G Ferrer S and Comin F 1996 *Phys. Rev. B***54** 8064
[8] Casiraghi C, Ferrari A and Robertson J 2005 *Phys. Rev.* B **72** 085401
[9] Tinchev S, Nikolova R, Dyulgerska J, Danev G and Babeva Tz 2005 *Solar Energy Materials&Solar Cells* **86** 421

IOP Conf. Series: Journal of Physics: Conf. Series **780** (2017) 012011 doi:10.1088/1742-6596/780/1/012011

Investigation of the structure and thermal behaviour of polymer liquid crystal / single wall carbon nanotubes nanocomposite

G Exner[1], Y Marinov[2] and E Perez[3]

[1] Department of Experimental physics, Faculty of Physics, Plovdiv University "Paisii Hilendarski", 24 Tzar Asen str., Plovdiv, Bulgaria

[2] Georgi Nadjakov Institute of Solid State Physics, Bulgarian Academy of Sciences, 72 Tzarigradsko Chaussee Blvd., BG-1784 Sofia, Bulgaria

[3] Instituto de Ciencia y Tecnología de Polímeros, Juan de la Cierva 3, 28006-Madrid, Spain

E-mail: exner_ginka@yahoo.com

Abstract. In the present work, nanocomposite of thermotropic polymer liquid crystal poly(heptane-1,7-dyil biphenyl-4,4'-dicarboxilate) and single wall carbon nanotubes was investigated. Nanocomposite films were casted from solution blended polymer liquid crystal and nanotubes. The structure and thermal behaviour of the nanocomposite were investigated by means of X-ray scattering and differential scanning calorimetry. The results show that there are two phase transitions on cooling and a single one on subsequent heating for both the neat polymer liquid crystal and nanocomposite. Hence, the smectic order of the polymer liquid crystal as well as its monotropic behaviour are preserved in the nanocomposite. The isotropic melt - smectic transition temperature in the nanocomposite is several degrees higher and the enthalpy of this process is much lower, suggesting heterogeneous nucleation of this phase on the surface of the nanotubes. The temperature of crystal structure formation during further cooling decreases in the nanocomposite showing a stabilization effect of the nanotubes on the smectic phase. Judging from the smaller enthalpy of the smectic-crystal phase transition and the new crystalline peak in the X-ray scattering patterns of the nanocomposite one could suggest a new crystalline form formation and this crystalline phase coexistence with smectic phases at lower temperatures.

1. Introduction

Invention of one and two dimensional nanomaterials, such as nanotubes, nanofibers, fullerenes and graphene open the challenging opportunity of developing new functional materials with improved or altered physical properties. Point of great scientific interest has become especially carbon nanotubes, CNT in their single wall, SW, and multi wall, MW, types because of their excellent mechanical, thermal, electrical and optical properties. By adding them into polymer matrices as nanofillers, one may design new nanocomposites [1-3]. In these new class of materials, new properties and morphologies were established and in many cases appropriate technological applications are already found. Polymer nanocomposites were mainly designed by solution blending, melt processing, and in situ polymerization [1,3]. The weak point in most cases is still the inhomogeneous dispersion and strong CNT aggregation tendency. Attempts to orient and align CNT in polymer nanocomposites by infiltration, applying magnetic or electric field, shear force or mechanical stretching are widely probed but they still do not represent general and very effective way of achieving the goal. In many cases to overcome the low miscibility and to increase to polymer - CNT interactions CNT are modified by acid oxidation, plasma modification and radical reaction and adding functional groups [2] but such approach bears the risk of introducing number of defects in the original nanotube structure and this might alter the initial nanotube properties.

In the last decades, some attempts to explore the interaction between nanotubes and liquid crystals, LC, are also made [4-11]. In some cases, a layer of aligned nanotubes has been used to aligned and orient LC molecules in smectic [5] or nematic [6] phases on its surface. Some authors [7] also discuss that LC provide a distinctive environment for controlling the alignment CNT but CNT from the other side are important for the enhancement and fine-tuning of LC properties. In some cases, CNT may improve the thermal stability without altering the LC texture [8]. Gursel et al. [9] report to be able to tune the nematic LC molecules alignment by variation in the CNT concentration. The dispersion and possible alignment of CNT via magnetic flied in presence of LC has been also discussed [10]. In present, some authors report already some applications of LC- CNT blends [11].

Even though, the polymer liquid crystals, PLC, combine the excellent mechanical properties with the strong tendency of self-organization in variety of different mesophases giving rise to excellent anisotropic physical properties, the possibility to combine them with nanofiller is just in its infancy and the number of scientific publications on this topic is very limited [12,13]. Such nanomaterials will have the additional advantage of easy processing and convenient molecular tailoring. PLC might be used in both respects to disperse/aligned the nanotubes and/or to influenced and to give us the way to tunes the PLC nanocomposite properties. The aim of the present work is to contribute to the investigations on polymer liquid crystal / carbon nanotube nanocomposites.

2. Experimental part

The nanocomposite was prepared by solution blending. As a first step the PLC poly(heptane-1,7-dyil biphenyl-4,4'-dicarboxilate), P7MB, and the single wall carbon nanotubes were separately dispersed in chloroform. P7MB, which synthesis is discussed elsewhere [14] was dissolved into chloroform (CAS:67-66-3, *Rai-chim, Bulgaria*) in a concentration 21.35 g/ml, at room temperature. The solution was bath-sonicated (*RH52 H, Bandelin Sonorex, Germany*) for 30 mins. Separately, single wall carbon nanotubes, SWCNT, (EC1.0, *Meijo Nano Carbon Co., Japan*) were added to chloroform in a concentration 0.14 mg/ml. The solvent was chosen as it is a second good solvent for SWCNT [15] and it is a common one for P7MB as well. A homogeneous suspension was achieved after 40 minutes of tip-sonication (*UP100 H Ultrasonic processor, Hielscher, Germany*) followed by 4 hours of bath-sonication. The P7MB solution and SWCNT suspension were then mixed and again bath sonicated for 4 hours. From the final suspension, films were casted. The solvent was left to evaporate overnight in a vacuum oven, at room temperature. The nanocomposite films were then heated to 195 °C and kept at this temperature of about 10 min, in order to get an isotropic melt and then either slowly cooled down to room temperature or quenched to room temperature (20°C), which is below the glass transition temperature of the P7MB T_g = 41°C.

X-ray scattering experiments were performed at Deutsches Elektronen-Synchrotron DESY, PETRA III station, P12 line (*BioSAXS*). The experimental range of d-space values was from 0.37 to 7.2 nm ($q = d/2\pi = 0.08722 \div 16.8777\ nm^{-1}$). Neat P7MB, quenched nanocomposite as well as slowly cooled samples were used.

Differential scanning calorimetry (*DSC 204 F1 Phoenix, Netzsch, Germany*), DSC, was employed to investigate the thermal behaviour and the phase transitions of the nanocomposite and to compare them with those of the neat PLC. The DSC equipment was temperature and sensitivity calibrated with 5 different materials. Small samples with a mass of about 5 mg were used. The samples were sealed in a pierced aluminium pan and the experiments were performed under argon atmosphere. The temperature protocol for the neat P7MB and for its composite was exactly the same. All samples were subjected

IOP Conf. Series: Journal of Physics: Conf. Series **780**(2017) 012011 doi:10.1088/1742-6596/780/1/01

initially to temperature equilibration of about 30 mins, at 20°C then heated at a linear rate of 10 °C/min up to the isotropization temperature of 195 °C. In order to achieve complete homogenisation and to delete possible thermal prehistory the samples were kept at this temperature for 10 minutes. The melt was then cooled down to 0 °C by applying a linear cooling rate of 10 °C/min. In order to investigate the thermal stability of the nanocomposite, the same temperature protocol was repeated two time after allowing the sample to equilibrate at 20 °C for 30 minutes.

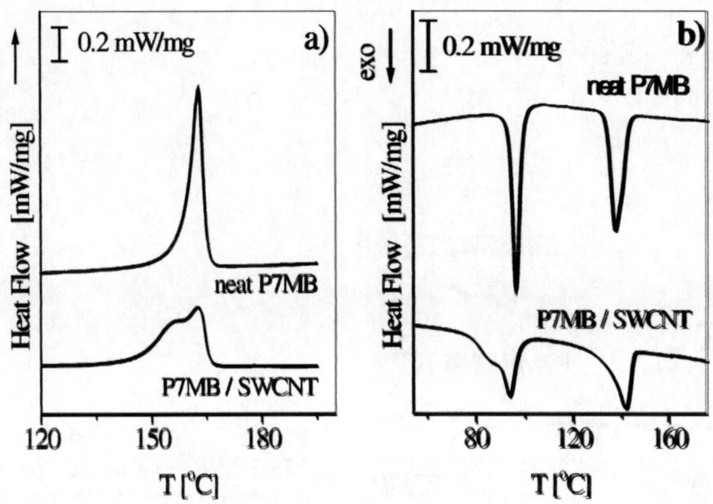

Figure 1. Differential scanning calorimetry thermograms of neat P7MB and nanocomposite P7MB / SWCNT: **a)** heating curves with melting endotherm at a heating rate 10 °C/min; **b)** cooling curves at a cooling rate 10 °C/min with two exotherms. The curves are translated vertically for better visualisation.

3. Results and discussion

The results from DSC show two phase transitions on cooling and single one on heating (figure 1) for the neat P7MB as well as for the nanocomposite of P7MB/SWCNT. Hence, the smectic order of P7MB is preserved in the nanocomposites as well as its monotropic behaviour. The transition temperatures of isotropic melt – smectic, I-Sm, for the nanocomposite ($T_{I-Sm}{}^{P7MB} = 141.7°C$) is several degrees higher than that of the neat polymer ($T_{I-Sm}{}^{P7MB/SWCNT} = 137.4°C$). In addition, the enthalpy of the neat polymer is $\Delta H_{I-Sm}{}^{P7MB} = 16.1\,J/g$ is much higher than that of the nanocomposite, where $\Delta H_{I-Sm}{}^{P7MB/SWCNT} = 11.7\,J/g$. The reason for the observed difference could be heterogeneous nucleation of the smectic phase on the surface of the nanotubes.

On further cooling, the smectic phase of the neat P7MB crystallizes [14], where the temperature of this transition as obtained from the DSC thermograms (Fig.1) is $T_{Sm-Cr}{}^{P7MB} = 96.1°C$ and enthalpy $\Delta H_{Sm-Cr}{}^{P7MB} = 16.3\,J/g$. It seems that the situation in the nanocomposite is similar and a crystallization takes place but here the transition is presented by two exotherms center at $T_{Sm-Cr,1}{}^{P7MB/SWCNT} = 93.7°C$ and $T_{Sm-Cr,2}{}^{P7MB/SWCNT} = 85.5°C$. The total enthalpy for this process is $\Delta H_{Sm-Cr}{}^{P7MB/SWCNT} = 13.8\,J/g$. The lower temperature suggests the stabilization effect of the nanotubes on the smectic phase. Interestingly, in both cases – quenching and slow cooling, a new crystalline peak appears (figure 2, marked with an arrow). This peak cannot be attributed to the

IOP Conf. Series: Journal of Physics: Conf. Series **780** (2017) 012011 doi:10.1088/1742-6596/780/1/012011

nanotubes only since they show a single peak, which still persists in the nanocomposite and its position is marked by a star in figure 2. Judging from the smaller enthalpy of the Sm-Cr phase transition and the new crystalline peak in the X-ray scattering patterns one would conclude that the nanofiller hinders the crystallization of the smectic phase and/or a new type of structure is formed. The mesogenic d-spacing for the neat P7MB is 1.73 nm, whereas for both different temperature treatments of the nanocomposite this spacing is 1.70 nm showing better order, expressed in existence of second order or the same d-spacing. The difference supports the idea of observing new crystalline form (or different smectic phase) in the presence of the nanofiller.

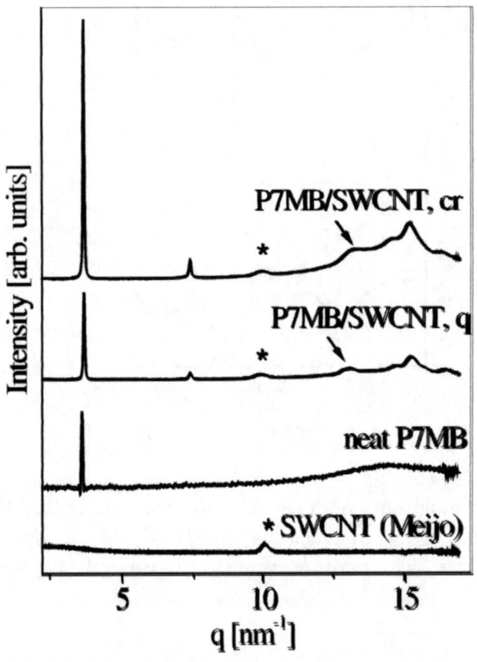

Figure 2. X-ray scattering curves of P7MB/SWCNT nanocomposite either quenched (q) or slowly cooled (cr) and those of neat P7MB as well as of SWCNT. The curves are translated vertically for better visualisation.

The melting temperature interval (figure 1) is almost the same for the neat polymer liquid crystal, having a maximum at ${T_{Cr-I}}^{P7MB} = 162.8°C$ and for its nanocomposite but the melting for the nanocomposite is represented by two endotherms with maxima ${T_{Cr-I,1}}^{P7MB} = 156.0°C$ and ${T_{Cr-I,2}}^{P7MB} = 162.6°C$. Comparing the values one can attribute the higher temperature peak to the crystalline phase melting whereas the lower temperature peak should be connected to the coexisting smectic phase.

The temperature behavior of the P7MB/ SWCNT nanocomposite is very reproducible, which was proven by several consecutive DSC runs, where the thermograms did not show any difference.

4. Conclusions

In the present work nanocomposite of PLC poly(heptane-1,7-dyil biphenyl-4,4'-dicarboxilate) and single wall carbon nanotubes were prepared by solution blending. The results show that in the nanocomposite the liquid crystallinity is preserved as well as its monotropic behaviour. The smectic phase appears by the mechanism of heterogeneous nucleation on the surface of the SWCNTs. The nanofiller has a stabilization effect on the smectic phase, shifting the smectic phase transition to higher temperature and crystallization of this smectic phase to lower temperatures. The low temperature phase is suggested to be a new crystalline form coexisting with the smectic phase, which could not undergo a phase transition. The nanocomposite displays temperature stability in the sense of complete reproducibility at subsequent heating and cooling cycles.

Acknowledgements
The authors acknowledge the INERA (EU project Research Potential FP7-316309-REGPOT-2012-2013-1) support for their participation at MELINA2016. This study was carried out as a part of Indo-Bulgarian joint research project DNTS/India 01/4 supported by the Ministry of Education and Science, National Science Fund of Bulgaria. The authors also thank BioSAXS beamline for their support and especially Dr. Vasyl Haramus.

References

[1] Goh P S, Ismail A F and Ng B C 2014 *Compos. Part A-Appl. S.* **56** 103–26
[2] Green M J, Behabtu N, Pasquali M and Adams W W 2009 *Polymer* **50** 4979–97
[3] Mittal G, Dhand V, Kyong Yop Rhee K Y, Soo-Jin Park S-J and Lee W R 2015 *J. Ind. Eng. Chem.* **21** 11–25
[4] Neeraj and Raina K K 2014 *Physica* B **434** 1–6
[5] Petrov M, Katranchev B, Rafailov P M, Naradikian H, Dettlaff-Weglikowska U and Keskinova E 2013 *J. Mol. Liq.* **180** 215–20
[6] Russell J M, Oh S, LaRue I, Zhou O and Samulski E T 2006 *Thin Solid Films* **509** 53 – 57
[7] Yadav S P and Singh S 2016 *Prog Mater Sci* **80** 38–76
[8] Chen X, Wu X, Zou J, Liu J and Chen J 2011 *Mater. Sci. Eng.* B **176** 425–30
[9] Gursel Y, Şenkal B F, Canli N Y, Özdemir Z G, Ahlatcioğlu E, Yılmaz Ö and Okutan M 2015 *Mat. Sci. Semicon. Proc.* **34** 182–88
[10] Yoo H J, Lee S Y, Youb N-H, Lee D S, Yeo H, Choi Y M, Goh M, Park J, Akagi K and Cho J W 2013 *Synthetic Met.* **181** 10– 17
[11] Lai Y-T, Chen Y-M, Liu T and Yang Y-J 2012 *Sensor Actuator A-Phys.* **177** 48 – 53
[12] Hu W-Q, Cui Z-K, Jin J, Dong Z-P, Li S-W, Wang P, Ding W and Li R 2011 *Appl. Surf. Sci.* **258** 507 – 12
[13] Ren Y, Wang T, Chen Z, Li J, Tian Q, Yang H and Xu Q 2016 *Appl. Surf. Sci.* **385** 521–528
[14] Pérez E, Todorova G, Krasteva M, Pereña J M, Bello A, Marugán M M and Shlouf M, 2003 *Macromol. Chem. Phys.* **204** 1791 – 99
[15] Bahr J L, Mickelson E T, Bronikowski M J, Smalley R E and Tour J M 2001 *Chem. Commun.* 193 – 94

IOP Conf. Series: Journal of Physics: Conf. Series **780** (2017) 012012 doi:10.1088/1742-6596/780/1/012012

The unique physical properties of the hydrogen bonded in dimers liquid crystals

M Petrov, B Katranchev and P M Rafailov

Institute of Solid State Physics, Bulgarian Academy of Sciences, 72 Tzarigradsko Chaussee Blvd., 1784 Sofia, Bulgaria.

E-mail: mpetrov@issp.bas.bg

Abstract. The dimerization of aromatic carboxylic acids, is the base of the structure formation of hydrogen bonded in dimers liquid crystals (HBDLCs), that exhibit non-conventional mesomorphism. The structural units of these LCs are amphiphilic-type molecules, which after suitable functionalization, induce supramolecular complexes, nanocomposites based on HBDLCs. The liquid crystalline character of the nanocomposites strongly dependent on intermolecular hydrogen bonds between symmetric, where the H-donors and H acceptors are contained in similar and non-symmetric HBDLCs, where the H-donors and H acceptors are contained in unlike molecules. The strength and non-covalent character of the hydrogen bonds provides both sufficient HBDLCs complex stability and bonding flexibility with a possibility to design and drive the supramolecular geometry. We investigated a series of nanocomposites produced by mixture of HBDLC (p-n-alkyloxybenzoic acid - nOBA), serving as matrices, with non-mesogenic (single walled carbon nanotubes - SWCNTs, hydroxypiridine – HOPY) and mesogenic (cholesteryl benzoate - ChB) nano-particles in various shapes and sizes. A set of new chiral ferroelectric phases were found in the nanocomposites, otherwise do not appearing in the pristine achiral HBDLCs materials. A molecular model of an unique low-temperature ferroelectric smectic phase C based on the molecular dimer ring symmetry reduction (bent dimer formation) towards to the lowest triclinic one is presented for both symmetric and nonsymmetric supramolecular liquid crystal complexes.

1. Introduction

The hydrogen bond (HB) enables various mesogenic and non-mesogenic compounds to form complexes exhibiting rich-phase polymorphism. It is an efficient alternative to covalent bonding as a way to stabilise liquid crystallinity. This bonding yields symmetric and non-symmetric complexes hydrogen bonded dimer liquid crystals (HBDLCs). HBDLCs are composed of a proton donor and acceptor molecules. Usually, for formation of HBDLCs, the carboxylic acids are preferable [1,2]. Thus, as part of a systematic development of this area the relationships between the phase behaviour and hydrogen bonding in a series of *p-n*-alkyloxybenzoic acids (nOBA) is an important LC topic.

The p-n-alkyloxybenzoic acid (nOBA) LCs are the basic components in forming supramolecular systems [3]. The strong ability to conformation of the dimeric structure in the principally achiral nOBAs creates the possibility for emergence of induced chirality [4]. The liquid crystalline nature of alkyloxybenzoic acids keeps at mixing with mesogenic or non-mesogenic compounds due to the hydrogen bonding. This non-covalent interaction predetermines the physical properties of nOBA LCs.

Due to the properties of the hydrogen bond, namely: directionality, reversibility and specificity, a wide range of homomeric and heteromeric complexes have been prepared exhibiting diversity of new electrooptical properties [4]. The non-symmetric supramolecular compound are complexes favoured by the stronger hydrogen bond between unlike units, while the symmetric ones exhibit such interaction between equivalent components. So, the formation of heteromeric complexes are mixtures of dissimilar H-donors and H-acceptors. A wide class of supramolecular liquid crystals was obtained by mixture of dimeric liquid crystals with set of non-mesogenic or mesogenic nanoparticles with different sizes and shapes [4]. In such produced composite states, the liquid crystal serves as a matrix with ability to impose its inherent high ordering to the bulk ordering of the mixture. Carbon nanotubes are among the most used as doping agent in the liquid crystal matrix [5-7].

The goal of the present work is to create both symmetric and non-symmetric supramolecular liquid crystals complexes, nanocomposites, where the dimeric nOBA liquid crystal serves like matrix. To produce the symmetric supramolecular liquid crystals complexes we use the dimeric LC 7OBA, mixed with SWCNTs and for non-symmetric nanocomosites we mix the dimeric LCs, 9OBA with mesogenic, cholesteryl benzoate (ChB) or non-mesogenic 4, hydroxypiridine (HOPY) nanoparticles, aiming to detect and explain the mechanism of induction of chirality and ferroelectricity, smectic C_G phase respectively, in the symmetric and non-symmetric nanocomposites. In order to understand the mechanism of the dimer ring bending, leading to symmetry decrease and C_G phase induction both in symmetric and non-symmetric complexes we present a balance of the intra and inter-molecular forces created by hydrogen bonding, and electrostatic (dipole moments) interactions.

2. Experimental results and discussion

2.1. For the symmetric nanocomposites we used 7OBA supplied by Sigma-Aldirch (Product of Japan) with strength of the linear hydrogen bond 20 kJ.mol^{-1}, which is two times bigger than those of 9OBA ($\approx$10 kJ mol^{-1}). The phase transitions temperatures of pristine 7,9OBA are:

$$\text{7OBA}\quad Cr \xleftrightarrow{92^o C} S_C \xleftrightarrow{98^o C} N \xleftrightarrow{146^o C} I$$

$$\text{9OBA}\quad Cr \xleftrightarrow{92^o C} S_C \xleftrightarrow{118^o C} N \xleftrightarrow{145^o C} I\,.$$

We examined the mixture 7OBA/SWCNTs with SWCNT concentration c = 0.01 wt% which was found to be optimal for the induction of chiral and ferroelectric states [8]. Purified SWCNT with diameters in the range 1.2–1.4 nm and an aspect ratio of 20–2000, assembled in thin bundles of >2 μm length, produced by arc discharge at Hanwha NanoTech Co., Ltd., were used as purchased. For the microtextural polarization analysis we used the known liquid crystal cell (LCC), two glass plates covered with ITO rubbed layer and the sample thickness d=8 μm. As was indicated in [8,9] the chiral induction in the nanocomposite was more effective due to both thin cell and good orientation in accordance with an optimal concentration. This effect determines the choice of these experimental parameters. We recall (see [8]) that at these experimental conditions the induced chiral and ferroelectric states at cooling are: I→N, N→N*, S_C^*→N_r^*, N_r^*→C_G, where N* - chiral nematic, S_C^* - ferroelectric smectic C, N_r^* - reentrant chiral nematic and C_G – low temperature smectic C_G state. We concentrate on the low-temperature smectic C_G state, which appears at 86°C at cooling. The microtexture of the produced smectic C_G state in the symmetric supramolecular nanocomplex 7OBA/SWCNT is indicated in figure 1. We remind that the nanocomposite, created in such way is symmetric, since the dimer ring binds two similar molecules (the two monomers) by two parallel linear hydrogen bonds.

The microtextural characteristics, equal left- and right-handed helices figure 1, are evidence for chirality and biaxiality of the LC mixture. Thus, the dimeric LC 7OBA/SWCNTs mixture, at cooling from I phase, exhibits consecutive local molecular symmetry reductions, leading to macroscopical one with the lowest triclinic symmetry C_1 (smectic C_G) where none of the principle axes of the second rank tensor, characterizing the orientational order, makes an angle of 0° or π/2 with the smectic layer's planes [10]. This strong local molecular symmetry decrease, can be realized only, when a local polarization induced by dimer conformation, or polar vector $\boldsymbol{P}$ bound up with the LC director $\boldsymbol{n}$, points out of the smectic layer's planes, thus the ground characteristic of the smectic C_G phase. Our microtexture and thermal analyses thus hint at a bent form of the dimer ring of 7OBA in the presence of the SWCNTs.

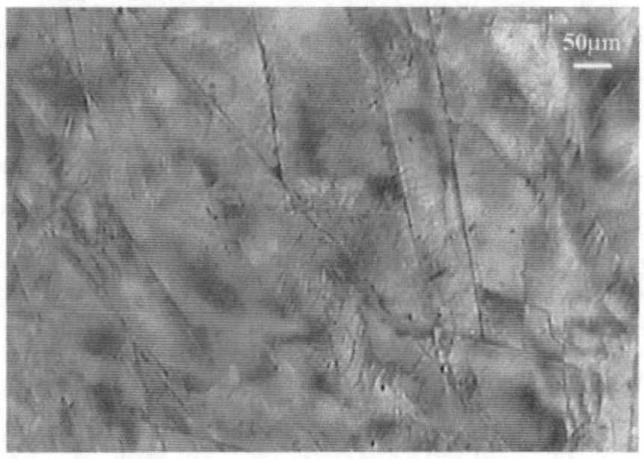

Figure 1. Colored mosaic C_G texture with equal left- and right-handed helices.

2.2. For the produce of the non-symmetric nanocomposites, we mix first the mesogenic nanocomponent ChB with 9OBA (9OBA/ChB) and second the non-mesogenic HOPY with 9OBA (9OBA/HOPY).

We recall that in the non-symmetric nanocomposite (non-symmetric supramolecular LC complex) the dimer ring of the dimeric LC is in the open state (open dimer) where the one free hydrogen bond, serving as H-bond donor, bind with free O atom (H-bond acceptor) of the nanoparticle. This means that the typical dimer symmetry, characterizing the pristine nOBA LC is destroyed and the symmetry of the complex is reduced. As we indicate in the figure 2 the smectic C_G phase is also induced in these cases, with textures strongly discernable from that of the symmetric 7OBA/SWCNTs complex. This difference hints for two different mechanisms for the C_G phase induction. Any more it indicates the unique role of the hydrogen bonding for build and design of nanocomposites.

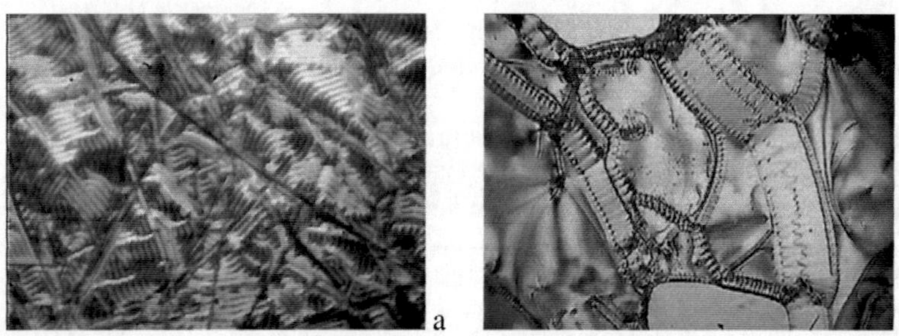

Figure 2. (a) The C_G texture of the nanocomposite 9OBA/ChB. (b) The C_G texture of the nanocomposite 9OBA/HOPY.

At this stage of discussion one notes, that the C_G microtextures of 9OBA/ChB and 9OBA/HOPY, to some extend, resemble those of the high-molecular bent-core C_G textures, which start to develop at cooling down to solid state, but without reaching a thermodynamical equilibrium. We designate such forming C_G phase as a developable. Contrary, the C_G phase texture of the symmetric nanocomposite 7OBA/SWCNTs was observed as stable and permanent with respect to the mosaic domains formation during the cooling down to the solid state. We designate such forming C_G phase as a developed.

In order to develop a model and corresponding mechanism of induction of C_G phase, both in symmetric and nonsymmetric supramolecular nanocomplexes, exposed by developed and developable C_G states, we use some characteristic for the pristine nOBA matrix parameters obtained on the base of the chemical AB INITIO MO calculation [11] and x-ray analysis [12]. The evolution of the hydrogen bonds of nOBA at cooling is indicated in [8,9]. Due to the thermal action, without external excitations, the symmetric dimer ring evolutes consequently to open dimer and oligomer (restricted polymer), but finally the system remains LC noncomposite. Thus in the combination with SWCNTs there is no chemical bond with external unsimilar molecule as that of the nanoparticle. For that case the characteristic geometrical and energetic parameter are as follows [11,12]: The geometrical parameters are for the monomer r inter-atom distances r: $r_{C=O}$=1.2202Å, r_{C-O}=1.3934Å, r_{O-H}=0.9896Å, angle $\gamma_{O-C=O}$=121.26° (angle between C=O and C–O groups); Closed Dimer: $r_{O...H}$=1.5024Å, $r_{C=O}$=1.2385Å, r_{C-O}=1.3521Å, r_{O-H}=1.0120 Å, angle $\gamma_{O-C=O}$=123.82°; Open Dimer: $r_{O...H}$=1.6761Å, $r_{C=O}$=1.2222Å, r_{C-O}=1.3798Å, r_{O-H}=0.9942Å, angle $\gamma_{O-C=O}$=122.71°. These parameters depict the disposition of the bonds participating in the vibrational spectra characteristic of the dimer ring. It is seen that there is a small variation of the geometrical parameters of the different molecular forms. Significant difference, however, in the energy and the dipole moment values exists when the three fundamental dimer forms appear. Thus, for 7OBA, the dipole moments μ in D (Debye) are 2.2643D for monomers, 0.0418D for the closed dimer and 4.7716D for the open dimers [11]. The dipole moment value of the closed dimer is approximately zero in accordance with simple logical considerations. In geometrical presentation, this means two approximately equal parallel dipoles with opposite directions. In the physical sense $\mu_{dim}\approx 0$ reflects the compensative role on μ of the closed dimer formation, implying that the molecule transforms from dipole (monomer) into quadrupole dielectric state. The calculated energies (ΔE) [11] of transition between monomer, closed dimer and open dimer of 7OBA or specially the energy of a linear hydrogen bond in the closed dimer is $1/2\Delta E_{mon\rightarrow dim}$ and is 24 kJ.mol^{-1} in good correspondence with that measured in [13].

The dimer structure $(R-CO-OH)_2$ (figure 3(a)) characterizes with 6 fundamental vibrations, which considered as Cartesian displacements, may be described as hydrogen bond stretching vibrations, in-plane or out-of- plane hydrogen bond bending vibrations or twisting vibrations [14,15]. Nevertheless, the closed dimer molecules are molecular quadrupoles. One can impose a dipole approximation, assuming the positive and the negative charges of the absorbing dimeric LC to segregate, following their distribution centers. In such way the ensemble of dimer molecules is a supra molecular LC system with an equivalent transition dipole moment $\boldsymbol{P}$, which can rotate around the C-C molecular axis. This transition dipole moment $\boldsymbol{P}$, or equivalent polar vector $\boldsymbol{P}$ expressing also the polar director, indicating the intra-molecular degree of freedom or conformational order parameter, deviates by an angle from the long molecular axis of the non-exited dimer as shown in figure 3. This figure presents the geometry of the polar vector deviation and depicts the dimer ring bending, quantified by the dimer bending kink vector $\boldsymbol{m} \sim \boldsymbol{n}_1\times\boldsymbol{n}_2$ where $\boldsymbol{n}_1$ and $\boldsymbol{n}_2$ are the unit vectors pointing along the two wings of the dimer. As a result, when bending of the dimer ring is present, $\boldsymbol{n}_1$ and $\boldsymbol{n}_2$ represent two local directors both including angle with the director $\boldsymbol{n}_o$ of the initially oriented dimeric LC system. Therefore, this is the geometry depicting the conformational order parameter in the dimeric system conformed by external actions.

Besides this important effect of the vibration softening hinting at dimer bending in the frame of the dimer ring structure, the Raman spectra in the range of 100–300 cm^{-1} (see [8,9]), also indicate π-stacking interactions between the SWCNT (carbon aromatic hexagon) and the liquid crystal molecule in its benzene core part. In the aspect of the competition between the forces promoted by these intra- and intermolecular interactions, and on the ground of the dominating bending (observed as s IR spectral polarization vibration [13]) one can conclude: that both perpendicular to the molecular optic axis forces are in coordination with the π-π electronic interaction, thus promoting a local molecular bending (bent dimer) and in consequence causing local and macroscopic symmetry decrease. Such local molecular symmetry decreases presume a macroscopic symmetry decrease characteristic for C_G. In the figure 3(b) a possible molecular model of the bulk interaction of the 7OBA LC matrix and the near-by SWCNT structure unit is depicted. H indicates the H-bonding vibration induced by the bending and detected by s IR light polarization. The direction of this s IR vibration is collinear with the stacking Raman vibration modulating the electronic π-π interaction between the LC benzenes and the carbon hexagons of the SWCNTs causing the LC dimer ring to bend and further to promote a multitude of sterically grazed bent dimers. As a result of this nanoscale process, a macroscopic electrical polarization $\boldsymbol{P}$ arises, which is characteristic of ferroelectric LC states [16,17].

IOP Conf. Series: Journal of Physics: Conf. Series **780** (2017) 012012 doi:10.1088/1742-6596/780/1/012012

We depict the two mechanisms and corresponding models for symmetric and non-symmetric supramolecular complexes in figure 3 and figure 4. For that purpose we use the above presented geometrical and energetic parameters and by torque balance of the non suppress forces we obtain the resultant, non zero space, vector bending the dimer ring, thus provoking the symmetry reduction to the lowest one triclinic characterizing the found by us smectic C_G phase (figure 3(c)). The polarization vector thus has two components: one is in the layer plane and the other one is an out-of-plane component.

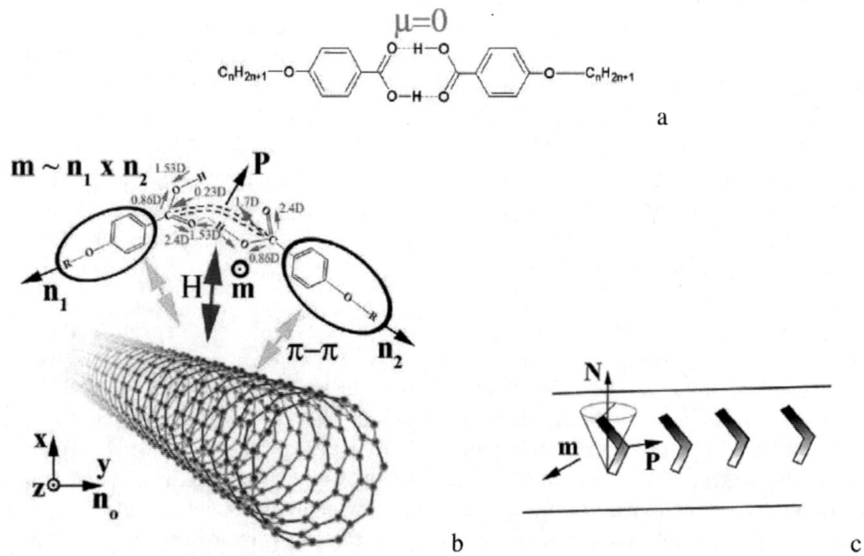

Figure 3. (a) The closed dimer. (b) A molecular model of the symmetric 7OBA/SWCNT nanocomposite. The average hydrogen bond (blue arrow), the π–π electronic interaction (green arrow) and the multitude of the acting local dipoles (red arrows), all participating in the force balance are indicated. The SWCNT hexagon point z axis. (c) The lowest one triclinic symmetry characterizing the smectic C_G phase.

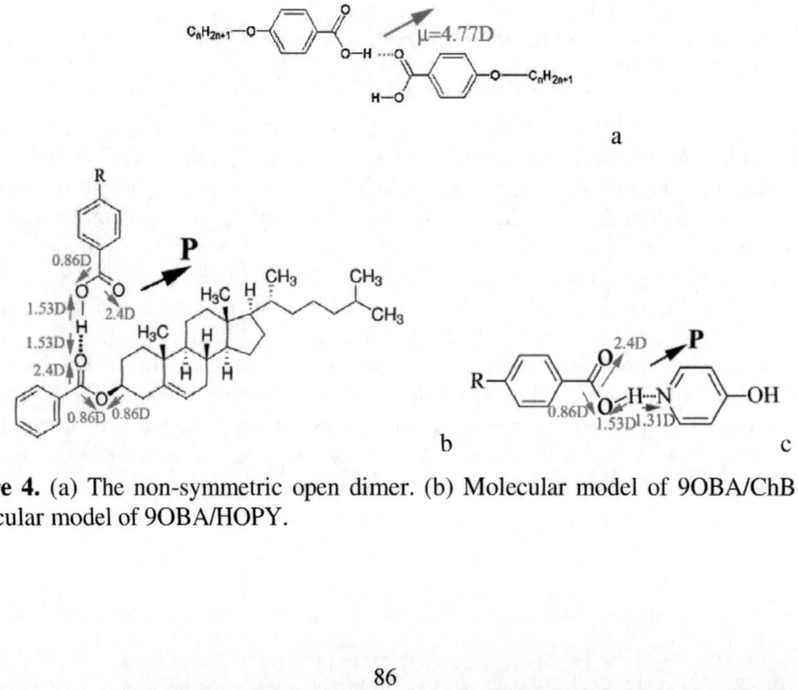

Figure 4. (a) The non-symmetric open dimer. (b) Molecular model of 9OBA/ChB (c). Molecular model of 9OBA/HOPY.

3. Conclusion

The symmetric and non-symmetric supramolecular liquid crystals, nanocomposites, both fulfilled on the base of the hydrogen bonding at mixing of dimeric nOBA liquid crystals with mesogenic or non-mesogenic nanoparticles, significantly differ with respect to generation and development of chiral ferroelectric states, including smectic C_G one. A possible reason for such a difference is that in the non-symmetric nanocomposites, exhibiting developable C_G state, the hydrogen bonding is heteromeric type, able to create non-homogeneous and non-stable in time dimer ring symmetry reduction, thus initiation thermodynamically unstable state. In this sense, the non-symmetric supramolecular liquid crystals, nanocomposites, resemble the high-molecular bent-core liquid crystal complexes. Unlike, in the symmetric supramolecular complexes, like that of 7OBA/SWCNTs, hydrogen bonding develops between two similar molecules, the two monomers of the nOBA molecule, within the electrical and steric field of the doped carbon nanotubes. The interaction of the two similar components of the nanocomposite 7OBA/SWCNTs, evidently leads to a thermodynamically stable developed C_G state. The future experimental and theoretical investigations are directed to further elaboration and confirmation of this model.

Acknowledgments
The work is supported by grant INERA EU project (FP7-316309-REGPOT-2012-2013-1) "Research and Innovation Capacity Strengthening of ISSP-BAS in Multifunctional Nanostructures" and Grant BK-04-14 from the Institute of Solid State Physics, Bulgarian Academy of Sciences.

References

[1] Kato T and Frechet J M J 2006 *Liquid Crystals* **33** 1429
[2] Paleos C M and Tsiourvas D 2001 *Liquid Crystals* **28** 1127
[3] Paterson D A, Martinez-Felipe A, Jansze S M, Marcelis A T M, Storey J M D and Imrie C T 2015 *Liquid Crystals* **42** 928
[4] Martinez-Felipe A and Imrie C T 2015 *J. Mol. Struct.,* **1100** 429
[5] Dierking I, Scalia G and Morales P 2005 *J Appl Phys.* **97** 044309
[6] Taylor J W, Kurihara L K and Martinez-Miranda L J 2012 *Appl Phys Lett.* **100** 173115
[7] Lynch M D and Patrick D L 2002 *Nano Lett.* **2** 1197
[8] Petrov M, Katranchev B, Rafailov P M, Naradikian H, Dettlaff-Weglikowska U, Keskinova E and Spassov T 2013 *Phys. Rev. E* **88** 042503
[9] Katranchev B and Petrov M 2016 *Phase Transitions* **89** 115
[10] de Gennes P G and Prost J 1993 *The Physics of Liquid Crystals* second edition (New York: Oxford University Press)
[11] Bobadova-Parvanova P, Parvanov V, Petrov M and Tsonev L 2000 *Cryst. Res. Technol.* **35** 1321
[12] Bryan R F and Fallon L 1975 *Journal of the Chemical Society-Perkin Transactions* **2** 1175
[13] Herbert A J 1967 *Trans. Faraday Soc.* **63** 555
[14] Petrov M, Antonova K, Kirov N, Tenev T and Ratajczak H 1994 *J. Mol. Struct.* **327** 265
[15] Antonova K, Katranchev B, Petrov M, Marcerou J P, Baran J and Ratajczak H 2004 *J Mol Struct.* **694** 105
[16] Jakli A, Kruerke D, Sawade H and Heppke G 2001 *Phys Rev Lett.* **86** 5715
[17] Chattham N, Korblova E, Shao R, Walba D M, Maclennan J E and Clark N A 2010 *Phys Rev Lett.* **104** 067801

IOP Conf. Series: Journal of Physics: Conf. Series **780** (2017) 012013 doi:10.1088/1742-6596/780/1/012013

Characterization of MAPLE deposited WO_3 thin films for electrochromic applications

S I Boyadjiev[1,2], N Stefan[3], I M Szilágyi[2,4], N Mihailescu[3], A Visan[3], I N Mihailescu[3], G E Stan[5], C Besleaga[5], M T Iliev[1,6], K A Gesheva[7]

[1]Institute of Solid State Physics, Bulgarian Academy of Sciences, 72 Tzarigradsko chaussee Blvd., 1784 Sofia, Bulgaria
[2] MTA-BME Technical Analytical Chemistry Research Group, Szent Gellért tér 4., H-1111, Budapest, Hungary
[3] National Institute for Lasers, Plasma and Radiation Physics, 409 Atomistilor Street, RO-77125, Magurele-Ilfov, Romania
[4] Budapest University of Technology and Economics, Department of Inorganic and Analytical Chemistry, 4 Műegyetem rakpart, H-1111, Budapest, Hungary
[5] National Institute of Materials Physics, 405A Atomistilor Street, Magurele-Ilfov, RO-077125, Romania
[6] Sofia University "St. Kliment Ohridski", 5 James Bourchier Blvd., 1164 Sofia, Bulgaria
[7] Central Laboratory of Solar Energy and New Energy Sources, Bulgarian Academy of Sciences, 72 Tzarigradsko chaussee Blvd., 1784 Sofia, Bulgaria

E-mail: boiajiev@gmail.com

Abstract. Tungsten trioxide (WO_3) is a widely studied material for electrochromic applications. The structure, morphology and optical properties of WO_3 thin films, grown by matrix assisted pulsed laser evaporation (MAPLE) from monoclinic WO_3 nano-sized particles, were investigated for their possible application as electrochromic layers. A KrF* excimer (λ=248 nm, ζ_{FWHM}=25 ns) laser source was used in all experiments. The MAPLE deposited WO_3 thin films were studied by atomic force microscopy (AFM), grazing incidence X-ray diffraction (GIXRD) and Fourier transform infrared spectroscopy (FTIR). Cyclic voltammetry measurements were also performed, and the coloring and bleaching were observed. The morpho-structural investigations disclosed the synthesis of single-phase monoclinic WO_3 films consisting of crystalline nano-grains embedded in an amorphous matrix. All thin films showed good electrochromic properties, thus validating application of the MAPLE deposition technique for the further development of electrochromic devices.

1. Introduction

Tungsten trioxide is studied for many applications, such as photocatalysis [1,2], sensing various toxic and inflammable gasses [3–7], water splitting [8] and solar energy conversion [9]. Also, WO_3 is the best known electrochromic material and its application in fields like large scale electrochromic devices as displays, smart windows and optical switching coatings is widely investigated [9–11].

Matrix Assisted Pulsed Laser Evaporation (MAPLE) technique has been validated for successful obtaining of thin films of sensitive materials avoiding thermal decomposition and irreversible degradation [12–15]. In MAPLE the target is frozen and consists of material to be ablated prepared as suspension or dissolved in an optically absorbing solvent with a high vapour pressure (*e.g.*, water, toluene). The purpose of the volatile solvent in the target is to aid desorption by absorbing most of the laser energy and vaporising when the laser energy is converted to thermal energy by photochemical processes, thus entraining the particles of interest to the substrate [16].

The expulsed substance is assembled on substrates where it forms a thin film with a thickness from a few to several hundreds of nm. The incident laser pulse used for MAPLE initiates two photo-thermal processes in the matrix: evaporation of the frozen composite target and transfer of the material onto the substrate [17]. Typically, the target consists of starting material (less than 10 wt.%) dissolved or suspended into a laser wavelength absorbing solvent when in frozen state. The material molecules reach sufficient kinetic energy by collective collisions with the evaporating solvent molecules so that a controlled transfer on the substrate in gas phase is ensured. Since the receiving substrate is usually kept at room temperature and the sticking coefficient of the solvent is nearly zero, the evaporated solvent is efficiently pumped away by the vacuum system. The solvent and solute concentration should be selected so that: the solute can be dissolved to form a dilute, particulate-free solution; most of the laser energy has to be absorbed by the solvent molecules rather than the solute ones; and no photochemical reaction is produced between solvent and solute [18]. By optimization of the MAPLE deposition parameters (solvent type, laser wavelength, repetition rate, laser fluence, solute concentration, substrate temperature, background gas and pressure), the process can proceed without significant material decomposition [19]. It was demonstrated that MAPLE could provide more crystalline layers as compared to PLD method from the same materials [12].

Highly transparent WO_3 thin films with as high as possible specific surface area are desirable for electrochromic applications, as the electrochromic properties depend not only on the nature and structure of the material, but also on its surface properties. Attainment of 5:1 ratios between the coloured and the bleached state (as required for many electrochromic devices), was preliminarily attempted here by such MAPLE.

2. Experimental

2.1. Materials and thin films preparation

Monoclinic WO_3 (m-WO_3) nano-sized particles were prepared by controlled annealing of hexagonal ammonium tungsten bronze, $(NH_4)_xWO_{3-y}$, in synthetic air at heating rate of 10 °C/min and purge gas flow rate of 130 mL/min [2]. The final annealing temperature was 600 °C. At temperatures over 550 °C WO_3 crystalizes predominantly monoclinic [20]. The MAPLE targets were fabricated using the as-synthesized m-WO_3 nano-sized particles.

The MAPLE set-up used in the experiments is presented schematically in figure 1. WO_3 thin films have been deposited using a KrF*excimer laser source (λ= 248nm, ζ_{FWHM} ≤25 ns), running at a repetition rate of 20 Hz. The substrates were ITO glass with sheet resistance of ~10 ohm/sq. (supplied by Sigma-Aldrich), as well as <110> single-crystalline Si wafers and glass for reference. Prior to deposition, all substrates were degreased in acetone for 30 min, washed in ethanol in an ultrasonic bath for 30 min, and then rinsed with deionised water.

IOP Conf. Series: Journal of Physics: Conf. Series **780** (2017) 012013 doi:10.1088/1742-6596/780/1/012013

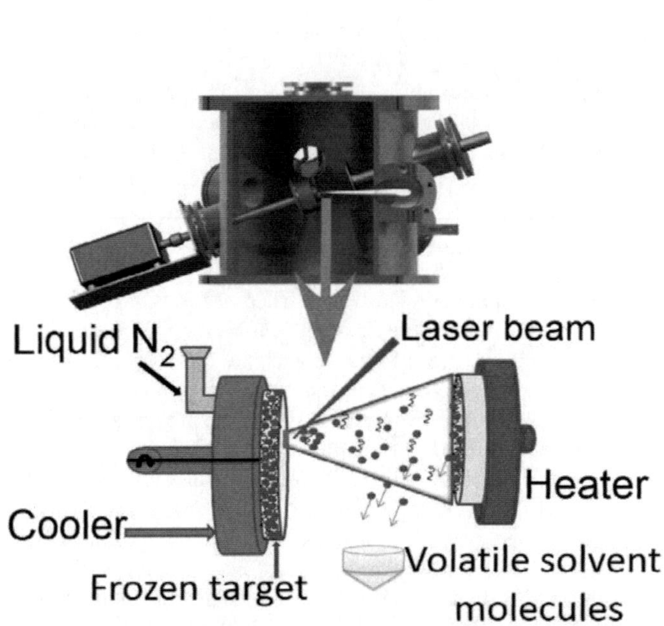

Figure 1. Typical MAPLE experimental set-up (schematics reproduced after Refs. 12 and 15).

The deposition parameters are summarized in table 1. The target-to-substrate distance ($D_{T\text{-}S}$), the oxygen pressure and the frequency of the ablation pulses (f) were constant for all samples. The deposition variables were: substrate temperature (T), number of pulses (consequently the thickness of the films), and fluence.

Table 1. Experimental conditions for WO_3 thin films deposited by MAPLE.

Sample code	$D_{T\text{-}S}$ (mm)	Fluence (J/cm^2)	f (Hz)	O_2 pressure (Pa)	T (°C)	No. Of pulses (K=10^3)
WO_3-251	40	1	20	16	250	12 K
WO_3-252		0.8			250	6 K
WO_3-253		0.6			250	4 K
WO_3-201		0.6			200	4 K

2.2. Characterization techniques

(*i*) The surface morphology of films was analysed by atomic force microscopy (AFM) in non-contact mode using an NT-MDT NTEGRA Probe NanoLaboratory system (NT-MDT NSG01 cantilever with tip radius of 10 nm).

(*ii*) The structure of the films was investigated by grazing incidence X-ray diffraction (GIXRD) using a Bruker D8 Advance diffractometer, in parallel beam setting, with Cu Kα (λ = 1.5418 Å) incident radiation. The incidence angle was set to 2°, and the scattered intensity was scanned within the range (20–60)° (2θ), with a step size of 0.04°, and 5 s per step.

(*iii*) Fourier transform infrared (FTIR) spectrometry studies were performed in transmission mode with a Shimadzu 8400S instrument. The investigated range was chosen between 500–4000 cm^{-1}, with a resolution of 4 cm^{-1}.

(*iv*) Cyclic voltammetry experiments were performed by a Bio-Logic Potentiostat/Galvanostat/ SP-200 in a standard three-electrode set-up. The cell used Pt as a counter electrode and a saturated calomel electrode as a reference electrode, altogether with the studied WO_3 films deposited onto ITO glass as working electrodes. The electrodes were immersed in electrolyte of 1 mol/l $LiClO_4$ dissolved in propylene carbonate.

3. Results and Discussion

FTIR spectrum of parent powder used for preparing the target for the MAPLE depositions is shown in figure 2. The IR characteristic vibration bands of monoclinic WO_3 were observed at the following wave numbers: 641 (ν(W–O) vibrational modes), 782 (WO_3 bending (δ) vibrations), 870 (WO_3 stretching (ν) vibrations) and 1010–1050 cm^{-1} (possible ν(W–O–W) vibrations) [21,22]. Supplemental bands positioned at (*i*) 1420 and (*ii*) 1630 cm^{-1} are ascribed to (*i*) stretching vibrations of hydroxyl groups (ν(OH)) which are strongly bonded to either water molecules or to surface oxygen atoms and/or of carbonate groups, and to (*ii*) in plane bending δ(HOH) of water molecules, respectively [22,23]. It has to be noted that the partial hydroxylation due to adsorption might negatively affect the long-term stability of such MAPLE deposited films used for electrochromic applications.

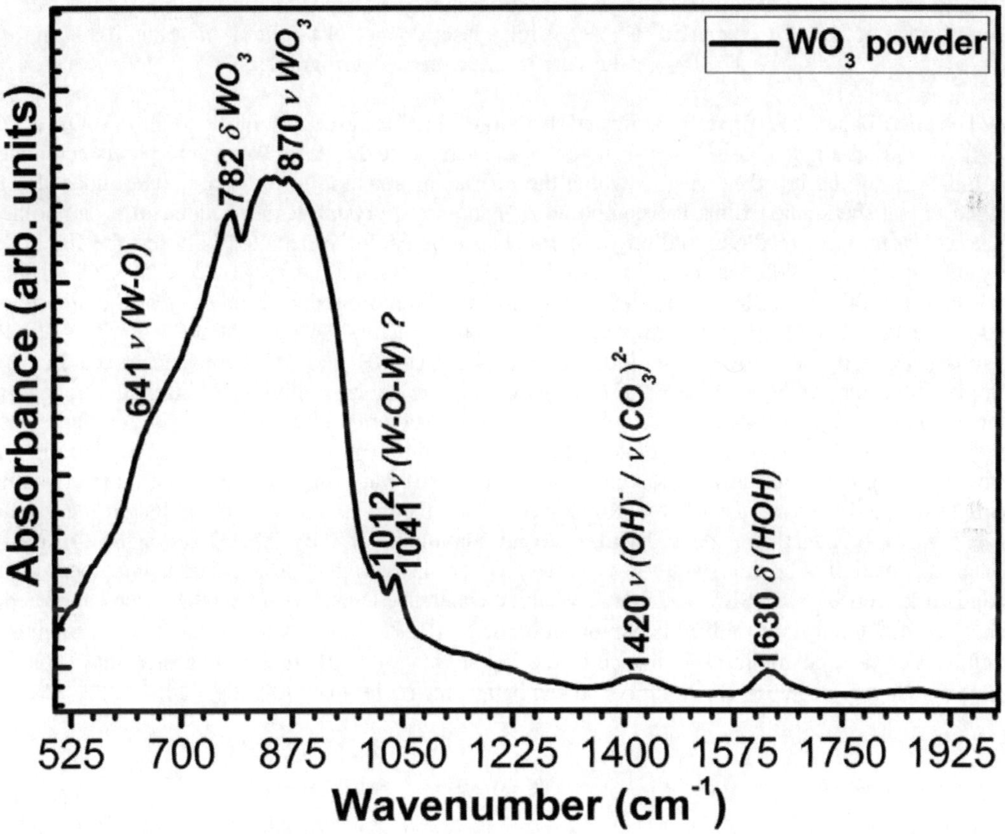

Figure 2. FTIR spectra of m-WO_3 powder used for preparing the target for the MAPLE depositions.

The morphology of the MAPLE deposited WO_3 films was studied by AFM on 0.4 μm x 0.4 μm areas. The typical morphology of films together with the root mean square roughness (R_{RMS}) are exemplified in figure 3 for three of the synthesized samples (*i.e.*, WO_3-201, WO_3-253, and WO_3-252). All films consisted of spheroidal grains with sizes in the range of about 10–50 nm. With the rise of the substrate temperature an increase of the larger grains population has been noticed (figure 3-bc), most probably due to coalescence processes. However, at a 200 °C (figure 3-a), the film surface appears more uniform and homogenous, which is indicative of a more well-ordered structure. The presence of the coalescence grains in the case of films deposited at 250 °C (figure 3-bc) results in an increase of the surface roughness from ~7 nm to ~12–13 nm.

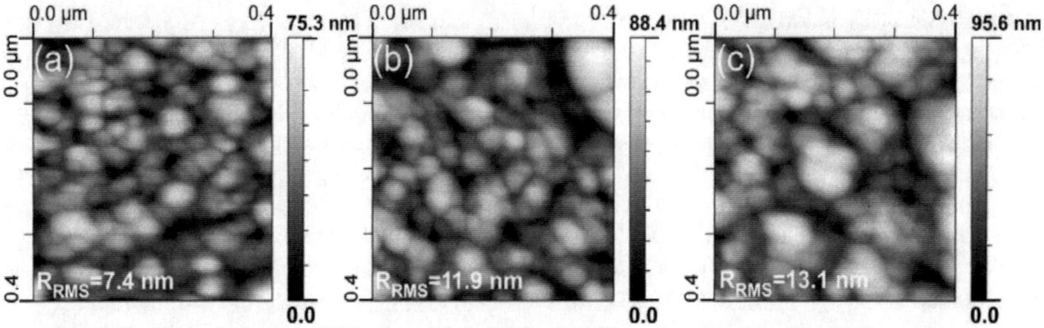

Figure 3. Characteristic AFM images of the WO_3 thin films deposited by MAPLE at substrate temperature of 200 °C (*a*) and 250 °C (*b,c*), using a laser fluence of 0.6 J/cm^2 (*a,b*) and 0.8 J/cm^2 (*c*). The R_{RMS} values are inserted on the micrographs.

The GIXRD patterns (figure 4) indicated that all MAPLE films consist of monoclinic WO_3 (ICDD: 01-083-0950) as single phase. One can note that the samples exhibited also an amorphous component as indicated by the broad humps on which the diffraction maxima of WO_3 are superimposed. This suggests that the studied films are composed of nano-sized crystallites embedded in an amorphous matrix. Differences in the crystallinity/amorphous ratios could be depicted between the films. The crystallinity of the WO_3 samples increased in the following order: 251<252<253<201. This is indicated by both the well-resolved WO_3 peaks and by the more reduced intensity of the amorphous halo. The laser fluence plays a prominent role on the structural features of the film. The GIXRD results suggested that a lower laser fluence (*e.g.*, samples 201 and 253 deposited at 0.6 J/cm^2) is propitious for the synthesis by MAPLE of films with improved crystallinity. Additionally, a moderate deposition temperature (*i.e.*, 200 °C) favors the overall structural quality of films (in good agreement with the AFM results), as at higher temperatures desorption phenomena can intensify during film growth, which can result in a (potentially) more faulty structure. In the case of the most crystalline sample (*i.e.* 201), having the most well-resolved peaks, the crystalline coherence length ("crystallite size") was estimated from the full-width at half maximum of the 200 diffraction line using the Scherrer equation. The line width was corrected for instrumental broadening using a corundum standard reference (NIST SRM 1976). A crystalline coherence length of ~26 nm has been calculated. The presence of a well-crystallized component in the MAPLE samples was considered encouraging, as could foster electrochromic performance of the films. Our previous studies suggested that WO_3 thin films with higher degree of crystallinity showed better electrochromic properties [24].

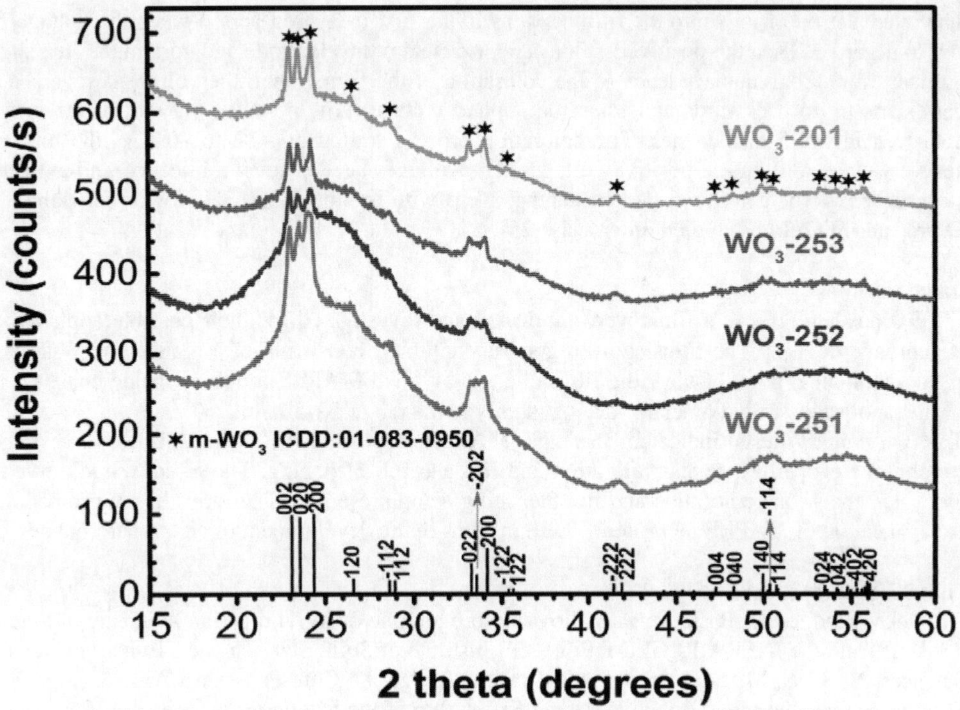

Figure 4. Comparative GIXRD patterns of MAPLE deposited WO_3 films.

Cyclic voltammetry measurements were performed on the WO_3 thin films fabricated by MAPLE, and the electrochromic effect was observed. The films showed strong coloration (the real value to be determined in further experiments), and fast (in a few seconds at the scanning rate of 20 mV/s) and full bleaching was observed. The effect was reversible for many cycles at comparably low voltages in the range –0.7/+0.4 V. For these MAPLE deposited WO_3 films, which are with higher specific surface, it was observed stronger coloration with respect to previously studied films deposited by magnetron sputtering or chemical vapour deposition (CVD). A characteristic cyclic voltammetry curve for 20 cycles measured at scanning rate of 20 mV/s is presented in figure 5.

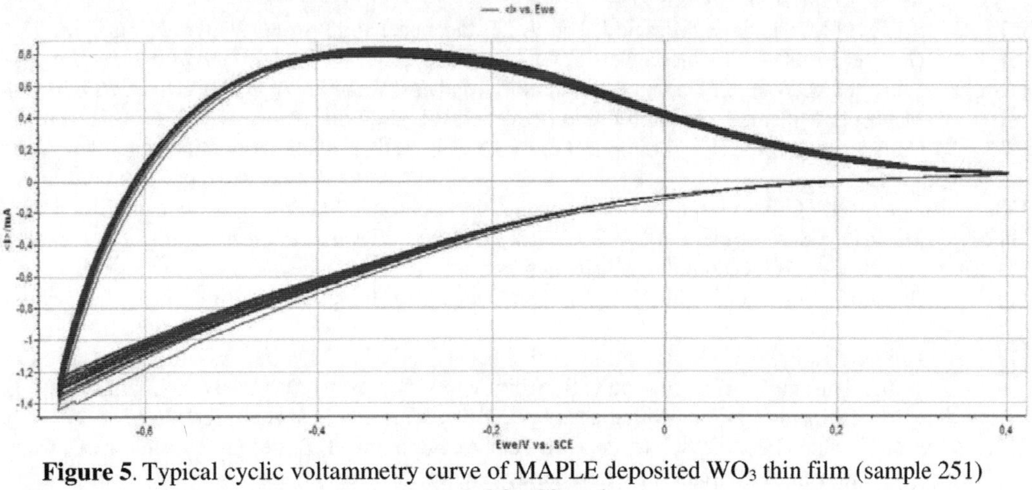

Figure 5. Typical cyclic voltammetry curve of MAPLE deposited WO_3 thin film (sample 251) measured for 20 cycles.

The intercalation of ions into the film happens in the first measured period when the voltage goes from 0 to negative. Then the positive Li^+ ions are attracted to the electrode and go into the free sites of the lattice, which consequently leads to the coloration of the film. When the voltage is reversed, the process starts to go backwards and when the applied voltage turns positive, big part of the ions de-intercalate, and the film bleaches. The coloration process started at –0.2 to –0.3 V, the bleaching started very close to 0, in the positive voltages. The voltages needed for ion injection and extraction, and consequently for coloration and bleaching, were with around 0.2–0.3 V lower compared with CVD prepared WO_3 films studied previously [25].

4. Conclusions

MAPLE deposited WO_3 thin films were fabricated and investigated for their possible application in electrochromic devices. The films showed good crystallinity, consisting of a monoclinic WO_3 phase. From morphological point of view the films elicited a typical MAPLE topology, being constituted of spheroid nano-grains with diameters varying mainly in the range 10–50 nm.

The preliminary tests indicated that the MAPLE deposited WO_3 thin films possess good electrochromic properties: strong coloration and fast and full bleaching. The effect was observed for many cycles, the strong coloration and full bleaching being preserved. These results are promising for future application of MAPLE deposited WO_3 thin films in the development of electrochromic devices.

Acknowledgements

S. I. B. acknowledges the Postdoctoral Fellowship programme of the Hungarian Academy of Sciences (2013–2015) and the INERA REGPOT Project of Institute of Solid State Physics, Bulgarian Academy of Sciences. N. S., N. M., A. V. and I. N. M. acknowledge the Core Program LAPLACE–IV 2016–2017. I. M. S. thanks a János Bolyai Research Fellowship of the Hungarian Academy of Sciences and an OTKA–PD–109129 grant. G. E. S. and C. B. thank the NIMP Core Programme PN 16 48–3/2016.

References

[1] Wicaksana Y, Liu S, Scott J and Amal R 2014 *Molecules* **19** 17747

[2] Szilágyi I M, Forizs B, Rosseler O, Szegedi A, Nemeth P, Kiraly P, Tarkanyi G, Vajna B, Var; Josepovits K, Laszlo K, Tóth A L, Baranyai P and Leskelä M 2012 *J. Catal.* **294** 119

[3] Blo M, Carotta M C, Galliera S, Gherardi S, Giberti A, Guidi V, Malagù C, Martinelli G, Sacerdoti Vendemiati B and Zanni A 2004 *Sens. Actuator B-Chem.* **103** 213

[4] Balázsi C, Wang L, Zayim E O, Szilágyi I M, Sedlacková K, Pfeifer J, Tóth A L and Gouma P-I 2C *J. Eur. Ceram. Soc.* **28** 913

[5] Szilágyi I M, Saukko S, Mizseic J, Tóth A L, Madarász J and Pokol G 2010 *Solid State Sci.* **12** 1857

[6] Guatwei L, Xian-ju T, Choongyew C, Gaiktin A and Zailani A B M 2011 *J. Mater. Sci. Eng.* **5** 27

[7] Szilágyi I M, Wang L, Gouma P-I, Balázsi C, Madarász J and Pokol G 2009 *Mater. Res. Bull.* **44** 50

[8] Marsen B, Miller E L, Paluselli D and Rocheleau R E 2007 *Int. J. Hydrogen Energy* **32** 3110

[9] Gesheva K A 2007 Thin film optical coatings for effective solar energy utilization (New York: Nc Science)

[10] Granqvist C G 1995 Handbook of Inorganic Electrochromic Materials (Amsterdam: Elsevier)

[11] Monk P M S, Mortimer R J and Rosseinsky D R 2007 *Electrochromism and Electrochromic Devi* (Cambridge: Cambridge university press)

[12] Boanini E, Torricelli P, Fini M, Sima F, Serban N, Mihailescu I N and Bigi A 2012 *J. Inorg. Bioche* **107** 65

[13] Jankovic A, Erakovic S, Ristoscu C, Mihailescu (Serban) N, Duta L, Visan A, Stan G E, Popa A C, Husanu M A, Luculescu C R, Srdic, V V, Janackovic Dj, Miskovic-Stankovic V, Bleot C, Chifiriuc M C and Mihailescu I N 2015 *J. Mater. Sci.–Mater. Med.* **26** 17

[14] Visan A, Stan G E, Ristoscu C, Popescu-Pelin G, Sopronyi M, Besleaga C, Luculescu C, Chifiriuc C, Hussien M D, Marsan O, Kergourlay E, Grossin D, Brouillet F and Mihailescu I N 2016 *Int. Pharmaceut.* **511** 505

[15] Visan A, Grossin D, Stefan N, Duta L, Miroiu F M, Stan G E, Sopronyi M, Luculescu C, Freche Marsan O, Charvilat C, Ciuca S and Mihailescu I N 2014 *Mater. Sci. Eng. B* **181** 56
[16] Mercado A L, Almond C E, Hoekstra J G and Fitzgerald J M 2005 *Appl. Phys. A* **81** 591
[17] Caricato A P and Luches A 2011 *Appl. Phys. A* **105** 565
[18] Mihailescu I N, Bigi A, Gyorgy E, Ristoscu C, Sima F and Oner E T2014 Biomaterial thin film by soft pulsed laser technologies for biomedical applications *Lasers in Materials Science (Springer Series in Materials Science vol 191)* eds M Castillejo, P M Ossi and L Zhigilei (Berlin: Springer) chapter 11 pp 271–294
[19] Mihailescu I N, Ristoscu C, Bigi A and Mayer I 2010 Advanced biomimetic implants based nanostructured coatings synthesized by pulsed laser technologies *Laser-Surface Interactions New Materials Production Tailoring Structure and Properties* (Springer Series in Materials Scien vol 130) eds A Miotello and P M Ossi (Berlin: Springer) chapter 10 pp 235–260
[20] Szilágyi I M, Sajó I, Király P, Tárkányi G, Tóth A L, Szabó A, Varga-Josepovits K, Madarász J a Pokol G 2009 *J. Therm. Anal. Calorim.* **98** 707
[21] Burdis M S, Siddle J R and Taylor S 1994 Proc. SPIE 2255, Optical Materials Technology for Ener Efficiency and Solar Energy Conversion XIII, 371 (September 9, 1994); doi:10.1117/12.185380
[22] Sharma N, Deepa D, Varshney P and Agnihotry S A 2001 *Thin Solid Films* **401** 45
[23] Diaz-Reyes J, Dorantes-Garcia V, Perez-Benitez A and Balderas-Lopez J A 2008 *Superficies y Va* **21** 12
[24] Boiadjiev S, Ivanova T, Rassovska M and Gesheva K A 2007 *J. Optoelectron. Adv. Mater.* **9** 264
[25] Gesheva K A, Ivanova T, Marsen B, Zollo G and Kalitzova M 2008 *J. Cryst. Growth* **310** 2103

IOP Conf. Series: Journal of Physics: Conf. Series **780** (2017) 012014 doi:10.1088/1742-6596/780/1/012014

Ab-Initio Molecular Dynamics Simulation of Graphene Sheet

S Kolev, I Balchev, K Cvetkov, S Tinchev and T Milenov

Institute of Electronics- Bulgarian Academy of Sciences, 72 Tzarigradsko Chausee Blvd., 1784 Sofia, Bulgaria

E-mail: kolevskk@abv.bg

Abstract. The study of graphene is important because it is a promising material for a variety of applications in the electronic industry. In the present work, the properties of a 2D periodic graphene sheet are studied with the use of ab initio molecular dynamics. DFT in the generalized gradient approximation is used in order to carry out the dynamical simulations. The PBE functional and DZVP-MOLOPT basis set are implemented in the CP2K/Quickstep package. A periodic box, consisting of 288 carbon atoms is chosen for the simulations. After geometry optimization it has dimensions 2964 x 2964 x 1500 pm and form angles of 90, 90, 60 degrees. The dynamical simulation is run for 1 ps in the NPT ensemble, at temperature $T = 298.15$ K. The radial distribution function shows a first peak at 142 pm, marking the bond length between carbon atoms. The density of states for the periodic systems is simulated as occupied orbitals represent the valence band and unoccupied ones the conduction band. The calculated bandgap, as expected is close to 0 eV.

1. Introduction

The study of graphene is important as it is a promising new material for a variety of applications in the electronic industry, as well as a highly durable or low resistant coating and medical applications [1]. Theoretical and computational approaches in studying conjugated bonds' surfaces such as graphene include mainly classical molecular dynamics and *ab initio* methods such as Density Functional Theory (DFT) and Post-Hartree-Fock methods. Classical molecular dynamics allow for exploring larger systems, but generally suffer in describing the electronic structure of the material as it relies on predefined force fields and potentials to represent bonding between atoms. The phonon properties of graphene have been investigated using classical molecular dynamics with Tersoff-2010, LCBOP28 and AIREBO29 potentials [2]. Molecular dynamics also allows the exploration of hybrid systems where graphene interacts with polymers [3].

Electronic effects in graphene are theoretically studied using *ab initio* static geometry optimization methods that allow electronic structure calculations. Density Functional Theory in generalized gradient approximation (GGA) is employed in exploring doped and defected structures [4]. Further, *ab initio* molecular dynamics offers a serious advantage over static geometry optimization methods: it allows time dependent investigation of the system's properties. For example, the Born– Oppenheimer molecular dynamics with GGA functional was used to study the properties of a graphene sheet using 60 carbon atoms in a periodic system and the most important structural factors were revealed, even with such a relatively small system [5].

As we began a program for a thorough study of ordered and disordered carbon including simulations of the density of states, vibrational frequencies etc., we aimed at several goals: (i) to conduct *ab initio* simulations using a large cell, (ii) to optimize the interatomic distances in the elementary cell and (iii) to verify our results for the molecular orbitals of graphene and compare them with experimental data. Here we present some preliminary results of *ab initio* simulations using a larger periodic system consisting of 288 carbon atoms.

2. Computational details

Ab Initio Born - Oppenheimer molecular dynamic simulations were performed using the CP2K/Quickstep package [6,7]. DFT was applied within GGA, using Perdew-Burke-Ernzerhof (PBE) functional [8]. The basis set DZVP-MOLOPT-SR-GTH [9], which is optimized for calculating molecular properties in gas and condensed phase, was applied for all atoms in the studied systems. For reducing computational cost Gaussian and Plane-Wave (GPW) method was used [10]. This method uses an atom-centered Gaussian-type basis to describe the wave functions and an auxiliary plane wave basis to describe the electron density [11]. It is worth noting that only the valence electrons are explicitly treated. Their interaction with the remaining ions is described using the pseudopotentials of Goedecker- Teter- Hutter (GTH) [12,13]. The charge density cutoff of the finest grid level is equal to 400 Ry. The number of used multigrids is 5.

All Born - Oppenheimer molecular dynamic simulations were carried out in the NPT ensemble with a timestep of 0.5 fs. The temperature was set to constant 298.15 K using a canonical sampling through velocity rescaling (CSVR) thermostat. Statistics were collected after thermal equilibration. The studied system consisted of a single graphene sheet represented by 288 carbon atoms.

3. Results and discussion

The elementary cell of graphene, used to prepare the periodic system is presented in figure 1. The triclinic cell consists of two carbon atoms and has dimensions (a = b) of 246 pm and the angle between C-C bonds is $\gamma = 120^0$. Such a triclinic cell is chosen over orthogonal as it permits for proper quantization of the Brillouin zone, proper description of permitted k-vectors and thus offers a realistic description of the structural and electronic properties of the system [14].

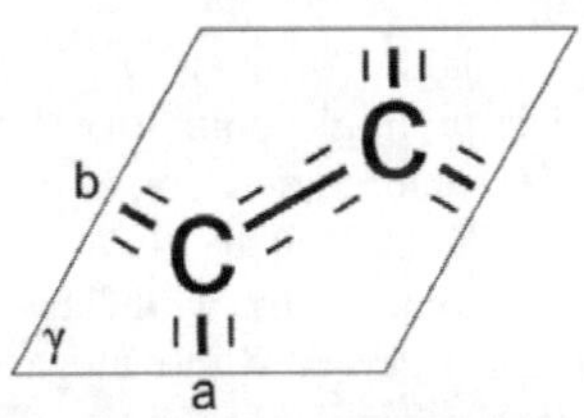

Figure 1. The elementary cell of graphene, used for the *ab initio* calculations.

The supercell used for dynamical simulations consists of a periodic system of 288 carbon atoms that occupy a hexagonal cell with dimensions A=12a and B=12b (i.e. 144 duplications of the elementary cell of graphene in the basis of the supercell) while the C- edge was fixed to 1500 pm. The volume along the C- edge above the basic graphene layer remains unoccupied. The dimension of the C- edge of the supercell is assumed to be long enough to eliminate the interaction between the graphene layers in the periodic system. The angles between A, B and C edges of the supercell are as follows: α- between A and C, β- between B and C and γ- between A and B with values of 90^0, 90^0 and 60^0, respectively. In order to prepare the initial system, geometry optimization is performed and the energy of the system is optimized taking into account the positions of the atoms as variables, e.g. the optimization is not performed on a structure with predefined overall dimensions but it is optimized computationally as well. After the geometry optimization, the dimensions obtained are: A= 2964 pm, B= 2964 pm and C= 1500 pm. The system should behave as a single layer. The actual system after optimization is presented in figure 2.

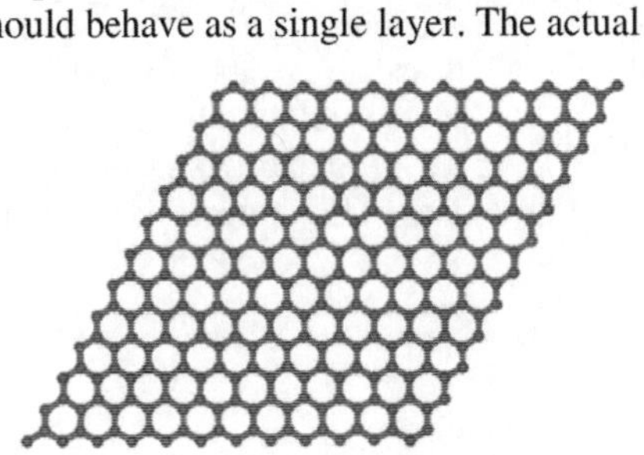

Figure 2. Periodic system consisting of 288 carbon atoms after the geometry optimization.

Born - Oppenheimer molecular dynamics of the optimized system was continued for 1 ps. The NPT ensemble was employed at pressure 1 atm, in order for the size of the system to be able to change with the vibrational and dynamical changes. Thermal equilibration is observed after 200 fs. After 1 ps, the radial distribution function (RDF) of one carbon atom versus the whole system was plot, the RDF is presented in figure 3. The first peak (at 142 pm) corresponds to the 3 neighboring C atoms of the first coordination shell, e.g. it is the actual C- C interatomic distance. The second peak at 246 pm corresponds to the 6 neighboring C atoms from the second coordination shell and the third peak at 286 pm corresponds to the 3 neighboring C atoms of the third coordination shell. The experimentally measured by transmission electron microscopy average C-C bond length in single-crystalline regions of single-layered graphene is also 142 pm, with a standard deviation of 5 pm [15,16].

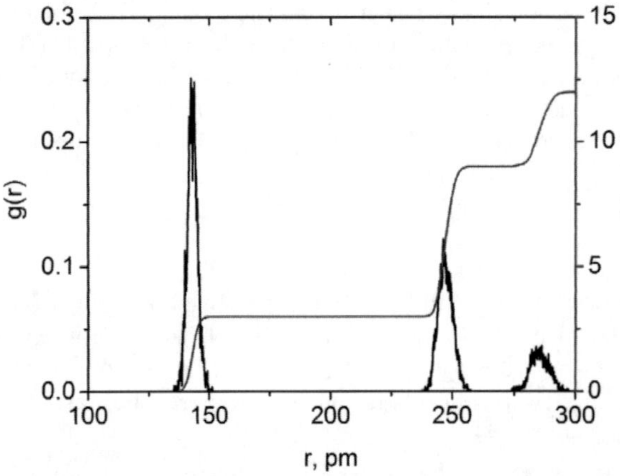

Figure 3. Radial distribution function (RDF) - black trace of one carbon atom versus the whole system and integral (blue trace) equal to the number of carbon atom neighbors.

The molecular orbitals of the graphene are calculated as a single point from the 250th fs of the simulation- figure 4. All occupied orbitals and the first 100 unoccupied are taken into account. The highest occupied molecular orbital (HOMO) – lowest unoccupied molecular orbital (LUMO) gap is 0.095 eV. The density of states is simulated as the molecule orbitals are presented with Gaussian peaks. A broad valence (occupied) and conduction (unoccupied) bandgaps are observed with bandgap of 0 eV- figure 5. The theoretical data [17] suggest that graphene does not have a bandgap as the conduction and valence bands coincide at the Dirac point.

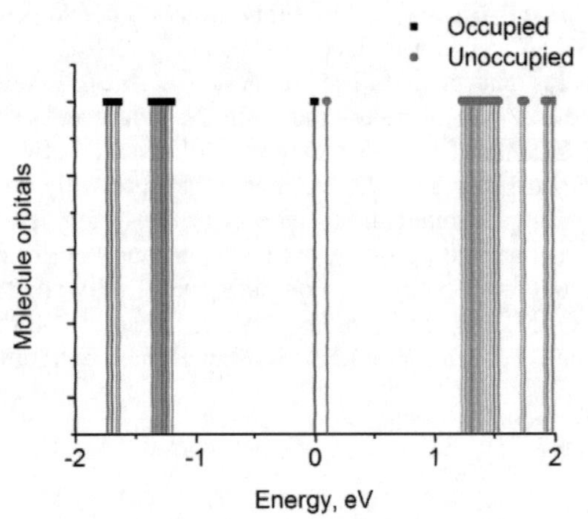

Figure 4. A plot of the molecule orbitals of the graphene layer. The occupied orbitals are presented as black lines and indicated by black squares while the unoccupied ones by red lines and indicated by red circles. The HOMO orbital is scaled to 0 eV.

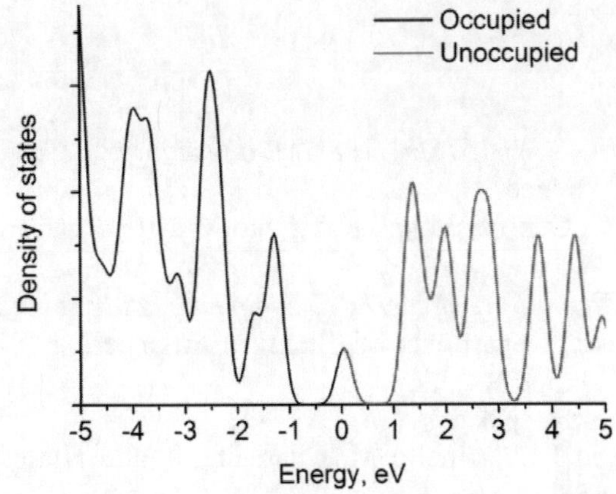

Figure 5. A plot of the density of states of the graphene layer. The valence band (occupied orbitals) and the conduction band (unoccupied orbitals) are presented by the black trace and the red trace, respectively. The HOMO orbital is scaled to 0 eV.

The properties of graphene depend on dimensions, intrinsic defects, even on the boundaries of the specimens studied. Ab initio calculations [18] show that energy gaps arise at the edges of graphene nanoribbons (GNR). Recently published experimental results show that: i) GNR can have a bandgap of about 1.5 eV and form heterojunctions [19] and ii) confirm that the zigzag edges of graphene nanoribbon open up a bandgap of 1.9 eV at the zigzag edge states [20]. The results obtained in our study, however, do not depend on the edge states effect, as periodic boundary conditions that treat the system as an infinite sheet with no edges are employed. Additionally, the system is sufficiently large (288 atoms) to represent correctly (in sufficient amount) the delocalized π and π* orbitals around HOMO and LUMO that form the valence and conduction bands.

4. Conclusions

The Born - Oppenheimer molecular dynamics with the Density Functional PBE represents correctly the structural properties of a single graphene layer and especially the C-C bond lengths. The optimized bond length is 1.42 pm which exactly coincides with the experimental findings. The electronic structure and density of states were also represented correctly and a bandgap close to 0 eV was obtained. The possibility to examine complex 2D systems with GGA Density Functional Theory provides a prospective tool for exploring functionalized or multilayered graphene as well as different sp^2/sp^3 mixed carbon phases.

Acknowledgements:

The authors gratefully acknowledge the financial support from the INERA Project as well as the COST Action MP 1204 "TERA-MIR Radiation: Materials, Generation, Detection and Applications".

References

[1] Ferrari A, Bonaccorso F, Fal'ko V, Novoselov K S, Roche S, Bøggild P, Borini S, Koppens F H L, Palermo V, Pugno N, Garrido J A, Sordan R, Bianco A, Ballerini L, Prato M, Lidorikis E, Kivioja J, Marinelli C, Ryhänen T, Morpurgo A, Coleman J N, Nicolosi V, Colombo L, Fert A, Garcia-Hernandez M, Bachtold A, Schneider G F, Guinea F, Dekker C, Barbone M, Sun Z, Galiotis C, Grigorenko A N, Konstantatos G, Kis A, Katsnelson M, Vandersypen L, Loiseau A, Morandi V, Neumaier D, Treossi E, Pellegrini V, Polini M, Tredicucci A, Williams G M, Hong B H, Ahn J H, Kim J M, Zirath H, van Wees B J, van der Zant H, Occhipinti L, Di Matteo A, Kinloch I A, Seyller T, Quesnel E, Feng X, Teo K, Rupesinghe N, Hakonen P, Neil S R T, Tannock Q, Löfwanderaq T and Kinaretba J 2015 *Nanoscale* **7** 4598

[2] Koukaras N, Kalosakas G, Galiotis C, Papagelis K 2015 *Sci. Rep.* **5** 12923

[3] Rissanou A, Bacova P, Harmandaris V 2015 *8th GRACM International Congress on Computational Mechanics.* Volos, Greece
[4] Hou Z, Wang X, Ikeda T, Terakura K, Oshima M and Kakimoto M 2013 *Phys. Rev. B* **87** 165401
[5] Bukowski B 2013 *A first principles study of graphene/TiO2 composite catalysts to improve photocatalytic activity*, B Sci Thesis, Department of Chemical Engineering, Worcester Polytechnic Institute, Worchester, United Kingdom
[6] The CP2K developers group 2012 http://www.cp2k.org/
[7] VandeVondele J, Krack M, Mohamed F, Parrinello M, Chassaing T and Hutter J 2005 *Comput. Phys. Commun.* **167** 103-128
[8] Perdew P, Burke K and Ernzerhof M 1996 *Phys. Rev. Lett.* **77** 3865-3868
[9] VandeVondele J and Hutter J 2007 *J. Chem. Phys.* **127** 114105
[10] Lippert G, Hutter J and Parrinello M 1997 *Molec. Phys.* **92** 477-487
[11] Lippert G, Hutter J and Parrinello M 1999 *Theor. Chem. Acc.* **103** 124-140
[12] Goedecker S, Teter M and Hutter J 1996 *Phys. Rev. B* **54** 1703-1710
[13] Hartwigsen C, Goedecker S and Hutter J 1998 *Phys. Rev. B* **58** 3641-3662
[14] Backe K 2014 *Functionalization of grapheme for biosensor-applications*, B Sci Thesis, Upsala Univeristy, Upsala, Sweden
[15] Rasool H I, Ophus C, Klug W, S Zettl and A Gimzewski J K 2013 *Nat. Commun.* **4** 2811
[16] Shevitski B, Mecklenburg M, Hubbard W A, White E R, Dawson B, Lodge M S, Ishigami M and Regan B C 2013 *Phys. Rev. B* **87**, 045417
[17] Cooper D R, D'Anjou B, Ghattamaneni N, Harack B, Hilke M, Horth A, Majlis N, Massicotte M and Vandsburger L Whiteway E Yu V 2012 *ISRN Condensed Matter Physics* 2012 5016861
[18] Son Y W, Cohen M L and Louie S G 2006 *Phys. Rev. Lett* **97** 216803
[19] Cai J, Pignedoli C A, Talirz L, Ruffieux P, Söde H, Liang L, Meunier V, Berger R, Li R, Feng X, Müllen K and Fasel R 2014 *Nat. Nanotechnol.* **9** 896
[20] Wang S, Talirz L, Carlo A. Pignedoli C A, Xinliang Feng X, Müllen K, Fasel R, and Ruffieux P 2016 *Nat. Commun.* **7** 11507

Impact of curvature on topological defects

L Mesarec[1], W Góźdź[2], A Iglič[1] and S Kralj[3,4]

[1] Laboratory of Biophysics, Faculty of Electrical Engineering, University of Ljubljana, Tržaška 25, 1000 Ljubljana, Slovenia
[2] Institute of Physical Chemistry, Polish Academy of Sciences, Kasprzaka 44/52, 01-224 Warsaw, Poland
[3] Condensed Matter Physics Department, Jožef Stefan Institute, Jamova 39, 1000 Ljubljana, Slovenia
[4] Faculty of Natural Sciences and Mathematics, University of Maribor, Koroška 160, 2000 Maribor, Slovenia

E-mail: samo.kralj@ijs.si

Abstract. We analyze the impact of *extrinsic* and *intrinsic* curvature on positions of topological defects (TDs) in two-dimensional (2D) nematic films. We demonstrate that both these curvature contributions are commonly present and are expected to be weighted by comparable elastic constants. A simple Landau-de Gennes approach in terms of tensor nematic order parameter is used to numerically demonstrate impact of the curvatures on position of TDs on 2D ellipsoidal nematic shells. In particular, in oblate ellipsoids the *extrinsic* and *intrinsic* elastic terms enforce conflicting tendencies to positions of TDs.

1. Introduction

Topological defects [1] (TDs) represent a popular intersciplinary subfield of science. For years, they are of hot interest in particle physics, condensed matter physics and even cosmology [2,3]. The reason for their popularity in all fields of science is their topological origin [1], which is blind for systems' microscopic details. In particular, there is growing evidence that *fields* represent basic entities of nature [4], and not "fundamental particles" as pictured by the Standard Model of nature. If this is the correct view than "fundamental particles" are emergent and correspond to relatively stable localized structures in a relevant *field.* Possible candidates for them are TDs as pioneering work of Skyrme suggests [5]. He has shown that pions, nuclear particles belonging to the meson family, could be represented as TDs in a relevant *field.* In the following years, mathematically related TDs, which are referred to as Skyrmions, have been detected and studied in more details in numerous different condensed matter systems [6,7,8] (e.g. in magnetism, liquid crystals...). Namely, due to interdisciplinary nature of TDs it is convenient to find experimentally adequate systems where static and dynamic fundamental properties of TDs could be relatively easily accessed [2].

The key property of TDs is their discrete topological charge q [2,3]. In general, topological charges play analogous role to electric charges: i.e., q of an isolated TD is conserved and a pair of TDs bearing opposite charge attract each other. Pairs of TDs bearing $\{q>0,-q\}$ are referred to as $\{$*defect,antidefect*$\}$ and have tendency to annihilate into a defect-less state. By contrast, TDs bearing the same sign of topological charges in general repel. In two-dimensions (2D) the topological charge is equivalent to the so called winding number m [3]. It quantifies the total rotation of the relevant orientational field divided by 2π on encircling the core of defect counterclockwise.

An adequate experimental testing bed to study TDs are liquid crystal (LC) phases [9]. In general, LC phases could be formed if they consist of relatively weakly interacting anisotropic molecules. For simplicity we henceforth restrict to thermotropic rod-like LC molecules, where LC ordering is stabilized in an adequate temperature interval. LC owe their experimental accessibility to unique combination of fluidity, softness, and optic anisotropy. Due to rich diversity of their configurations they can possess any topological defect configuration [2]. For instance, their structure could be relatively easily studied using optic polarizing microscopy. Nematic configurations represent simplest LC structures. In the nematic bulk phase LC molecules tend to be homogeneously aligned along a single symmetry breaking axis, while centers of masses of molecules flow roughly like in a fluid.

Effectively two-dimensional nematic LC structures are of particular interest. First of all, they are relatively easily accessible for mathematical and numerical investigations. Secondly, such structures are of interest for various applications, in particular those exhibiting TDs. For instance, nematic shells [10,11] consisting of supra micron-sized colloids covered with relatively thin LC films are of interest for various photonic applications. Furthermore, biological membranes are essentially thin LC films [12,13]. So far, in almost all theoretical attempts, covariant derivatives [13,14,15] were employed in expressing free energy elastic terms of 2D ordered systems. Such approaches took into account only the so called *intrinsic* curvature [16] elasticity which is associated with spatial variations of a relevant orientational field within the curved surface. It has been shown that in this case surface regions with high absolute value of the Gaussian curvature attract TDs [14,15,17,18]. Furthermore, electrostatic analogy works well where positive (negative) Gaussian curvature plays similar role as spread negative (positive) topological charge of TDs [14,18]. However, recently it has been demonstrated that *extrinsic* curvature [16,19,20] elasticity terms are in general also present and could have important impact on resulting structures. This elasticity takes into account that a curved 2D manifold is embedded in 3D geometry. In some cases a structure could simultaneously maximize and minimize *intrinsic* and *extrinsic* free energy costs contributions and vice versa [16,19].

In the present paper we demonstrate relative importance of *intrinsic* and *extrinsic* terms on orientational ordering within nematic shells exhibiting spherical topology. For simplicity we restrict to structures displaying cylindrical symmetry. We first demonstrate using a minimal model that both contributions are in general present. Then we numerically demonstrate cases, where *extrinsic* elasticity could play significant role.

2. *Intrinsic* and *extrinsic* curvature

We consider a thin nematic LC film and describe local orientation of LC molecules by the unit vector field $\vec{n}$ lying within a curved 2D film. We consider rod-like LC molecules, therefore the orientations $\pm\vec{n}$ are equivalent. For illustration purpose we chose the simplest possible free energy density expression $f = k|\nabla_s \vec{n}|^2$ [9]. Here k is the positive elastic constant, the operator $\nabla_s = (\underline{I} - \vec{\nu} \otimes \vec{\nu})\nabla$ is the surface gradient [17], $\underline{I}$ is the 3D identity tensor, ∇ is the convential 3D gradient operator, $\vec{\nu}$ is the surface normal, and $\otimes$ is the tensorial product.

It is convenient to parametrize $\vec{n}$ in the surface principal curvature frame $(\vec{e}_1, \vec{e}_2)$ in which the curvature tensor $\underline{C} = C_1 \vec{e}_1 \otimes \vec{e}_1 + C_2 \vec{e}_2 \otimes \vec{e}_2$ is diagonal, $\{C_1, C_2\}$ are the corresponding principal curvatures, and $\vec{\nu} = \vec{e}_1 \times \vec{e}_2$. Therefore

$$\vec{n} = \vec{e}_1 cos\vartheta + \vec{e}_2 sin\vartheta. \tag{1}$$

Taking into account equations [21]

$$\nabla_s\vec{n} = cos\vartheta\nabla_s\vec{e}_1 - sin\vartheta\vec{e}_1 \otimes \nabla_s\vartheta + sin\vartheta\nabla_s\vec{e}_2 + cos\vartheta\vec{e}_2 \otimes \nabla_s\vartheta, \tag{2a}$$

$$\nabla_s\vec{e}_1 = \kappa_{g1}\vec{e}_2 \otimes \vec{e}_1 + \kappa_{g2}\vec{e}_2 \otimes \vec{e}_2 - C_1\vec{v} \otimes \vec{e}_1, \tag{2b}$$

$$\nabla_s\vec{e}_2 = -\kappa_{g1}\vec{e}_1 \otimes \vec{e}_1 - \kappa_{g2}\vec{e}_1 \otimes \vec{e}_2 - C_2\vec{v} \otimes \vec{e}_2, \tag{2c}$$

it follows

$$|\nabla_s\vec{n}|^2 = \left|\nabla_s\vartheta + \vec{A}\right|^2 + \vec{n}.\underline{C}^2\vec{n}, \tag{3a}$$

$$\vec{A} = \kappa_{g1}\vec{e}_1 + \kappa_{g2}\vec{e}_2. \tag{3b}$$

Here $\{\kappa_{g1}, \kappa_{g2}\}$ are geodesic curvatures along $\{\vec{e}_1, \vec{e}_2\}$, $\vec{A}$ is the spin connection [22], which can be expressed in terms of the Gaussian curvature $K = C_1C_2$ as $K = \left|\nabla \times \vec{A}\right|$.

Therefore, the free energy density could be expressed as the sum $f = f^{(int)} + f^{(ext)}$ of *intrinsic* ($f^{(int)}$) and *extrinsic* ($f^{(ext)}$) elastic contributions, where

$$f^{(int)} = k\left|\nabla_s\vartheta + \vec{A}\right|^2, \tag{4a}$$

$$f^{(ext)} = k\vec{n}.\underline{C}^2\vec{n} = k(C_1cos^2\vartheta + C_2sin^2\vartheta). \tag{4b}$$

The *intrinsic* term enforces homogeneous orientational ordering for surfaces characterized by K=0. By contrast, surface patches with nonzero Gaussian curvature are geometrically frustrated. Consequently, spatially nonhomogenous ordering mimimizes $f^{(int)}$. More exactly, a local orientational structure obeys the so called parallel transport condition [21]:

$$\nabla_s\vec{n} = -(\vec{v} \otimes \underline{C})\,\vec{n}. \tag{5}$$

It originates from the incompatibility of straight-parallel direction on surfaces with $|K| > 0$. Using the parametrization Eq.(1) the condition yields $\nabla_s\vartheta = -\vec{A} = -(\kappa_{g1}\vec{e}_1 + \kappa_{g2}\vec{e}_2)$, and consequently $f^{(int)} = 0$. Note that the approaches emplying covariant derivatives yield only the *intrinsic* contribution.

The *extrinsic* term acts as an effective external field. Namely, it tends to align $\vec{n}$ along the principal direction exhibiting minimal curvature.

3. Numerical simulation

In order to demonstrate impact of *extrinsic* curvature we employ a simple Landau-de Gennes type model, where nematic orientational order is represented by the tensor order parameter $\underline{Q}$ [17,21].

3.1. Model

In the diagonal form it can be expressed as

$$\underline{Q} = \eta(\vec{n} \otimes \vec{n} - \vec{n}_\perp \otimes \vec{n}_\perp). \tag{6}$$

Here $\eta \in [0,1/2]$ and $-\eta$ are eigenvalues of eigenvectors $\vec{n}$ and $\vec{n}_\perp$, respectively. Localised points on 2D menifolds exhibiting nematic ordering in general reveal presence of TDs. We calculate the degree of nematic ordering by minimizing the relevant free energy of a system. We express the relevant dimensionless free energy density as [17,19,20]

$$f = \left(\frac{R}{\xi_0}\right)^2 \left(t\left|\underline{Q}^2\right| + \left|\underline{Q}^2\right|^2 + k_i\left|\nabla_s\underline{Q}\right|^2 + k_e\underline{Q}^2.\underline{C}\right). \tag{7}$$

Here R is a geometrically imposed length characterizing a system, ξ_0 estimates the order parameter length deep in the nematic phase, the quantity $t=(T-T_c)/T_c$ is the reduced temperature, T_c is the critical temperature below which spatially homogeneous nematic ordering is stable on a flat surface. The relative strength of {*intrinsic,extrinsic*} elastic contribution is weighted by dimensionless constants $\{k_i,k_e\}$, where we set k_i=1.

In simulations we consider ellipsoidal shapes which we parametrize in the Cartezian system (x,y,z) determined by the unit vectors $(\vec{e}_x, \vec{e}_y, \vec{e}_z)$ as

$$\vec{r} = b(\sin(\mathrm{v})\cos(\mathrm{u})\vec{e}_x + \sin(\mathrm{v})\sin(\mathrm{u})\vec{e}_y) + a\cos(\mathrm{v})\vec{e}_z. \tag{8}$$

The position vector $\vec{r}$ determines a point on an ellipsoidal surface, $\mathrm{v} \in [0,\pi], \mathrm{u} \in [0,2\pi]$, and R in Eq.(7) corresponds to the minimal length of the pair $\{a,b\}$. The ellipsoidal shells are prolate for $a/b>1$ and oblate for $a/b<1$. More details are given in [17].

3.2. Nematic textures

In figures 1 we plot a typical nematic structure on a spherical shell (a/b=1). In figure 1a we plot superimposed scalar order parameter η and the orientational field $\vec{n}$, where states $\pm\vec{n}$ are equivalent. The corresponding scalar order parameter graph in the (u,v) plane, which clearly reveals position of four TDs bearing topological charge m=1/2, is shown in figure 1b. The quantity η_c determines the equilibrium value of η in a flat film. In this case the *extrinsic* term is absent because both principal curvatures are equal. The TDs are placed in such a way to maximize their mutual separation, because TDs repeal each other.

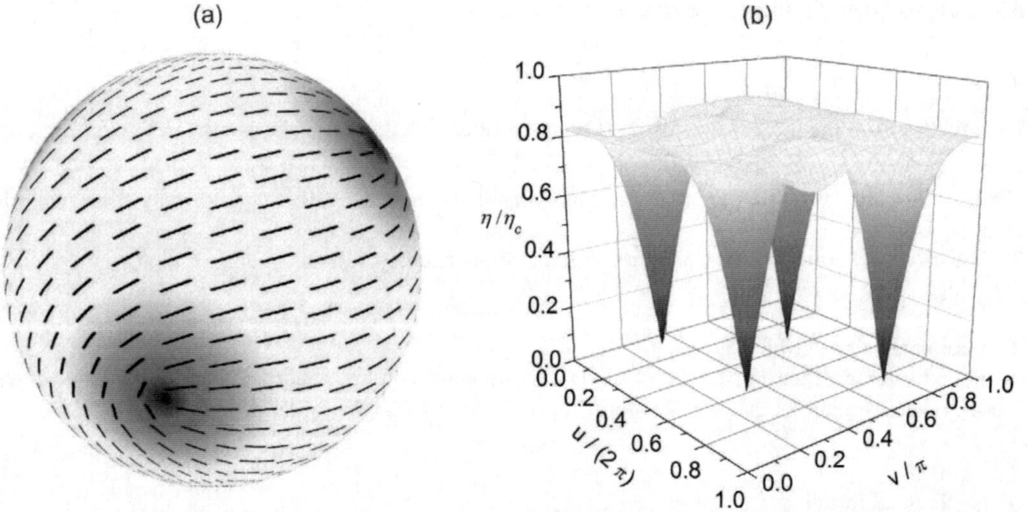

Figure 1. Typical nematic ordering on a spherical shell. The color code represents the $\boldsymbol{\eta/\eta_c}$ ratio. The panel (a) also shows the directions of molecules on a sphere. The panel (b) represents the calculated nematic texture in the $(\mathbf{u},\mathbf{v})$-plane. Parameters: $\boldsymbol{a/b=1}$, $\boldsymbol{R/\xi_0=20}$, $\boldsymbol{k_i=1}$, $\boldsymbol{k_e=0}$, $\boldsymbol{t=-0.03}$.

Next we deform spheres to form oblate and prolate ellipsoids. In figures 2 (nematic textures depicted on ellipsoidal shapes) and figures 3 (nematic textures in the (u,v) plane) we demonstrate impact of shape and extrinsic curvature on position of TDs. We first consider cases for k_e =0 for which the *extrinsic* elasticity is absent. A typical nematic texture for a prolate structure for is shown in figure 2a and figure 3a. TDs are pushed towards the poles because they are attracted by the positive Gaussian curvature, which exhibits maximum at the poles, see figure 4a. This attraction of TDs bearing positive topological charges with regions exhibiting $K>0$ competes with the mutual repulsion among TDs. Note that for strong enough prolateness pairs of $m=1/2$ TDs merge into $m=1$ TDs localized at poles of ellipsoids [17]. In oblate structures K exhibits maximum at the equatorial great circle as figure 4a demonstrates. In figure 2b and figure 3b we show a case where attraction between K>0 and TDs is strong enough to localize TDs at the equatorial circle.

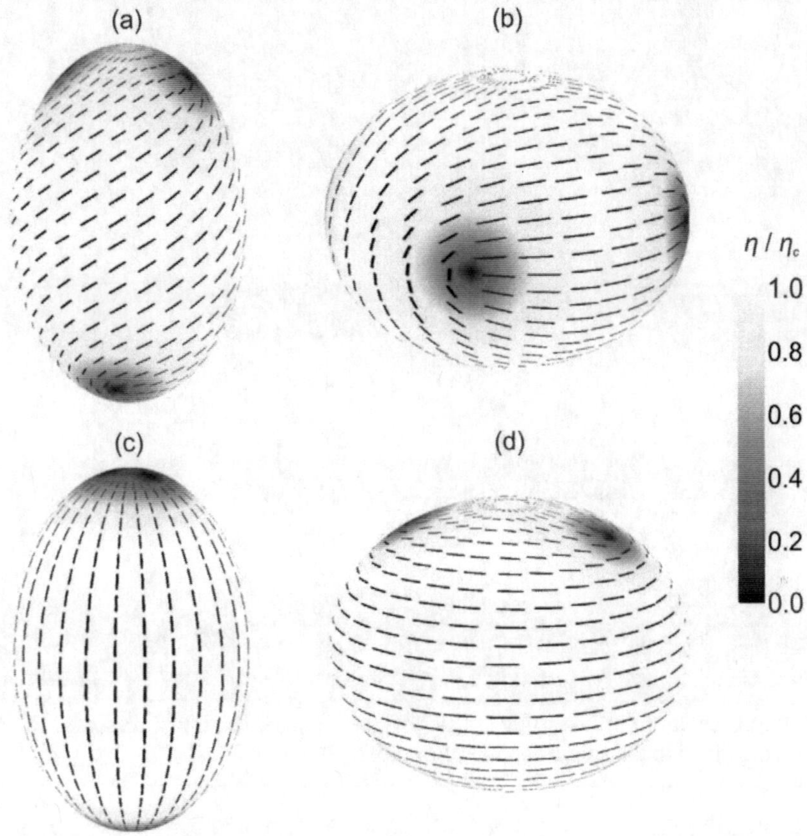

Figure 2. The degree of nematic ordering and the directions of molecules on prolate $\boldsymbol{a/b = 1.5}$ (a,c) and oblate $\boldsymbol{b/a = 1.28}$ (b,d) ellipsoids. The nematic textures in panels (c) and (d) show the impact of the *extrinsic* elastic contribution: $\boldsymbol{k_e = 1}$, while the configurations in panels (a) and (b) were calculated without the extrinsic term: $\boldsymbol{k_e = 0}$. Other parameters were set as follows: $\boldsymbol{R/\xi_0 = 20,\ t = -0.03}$.

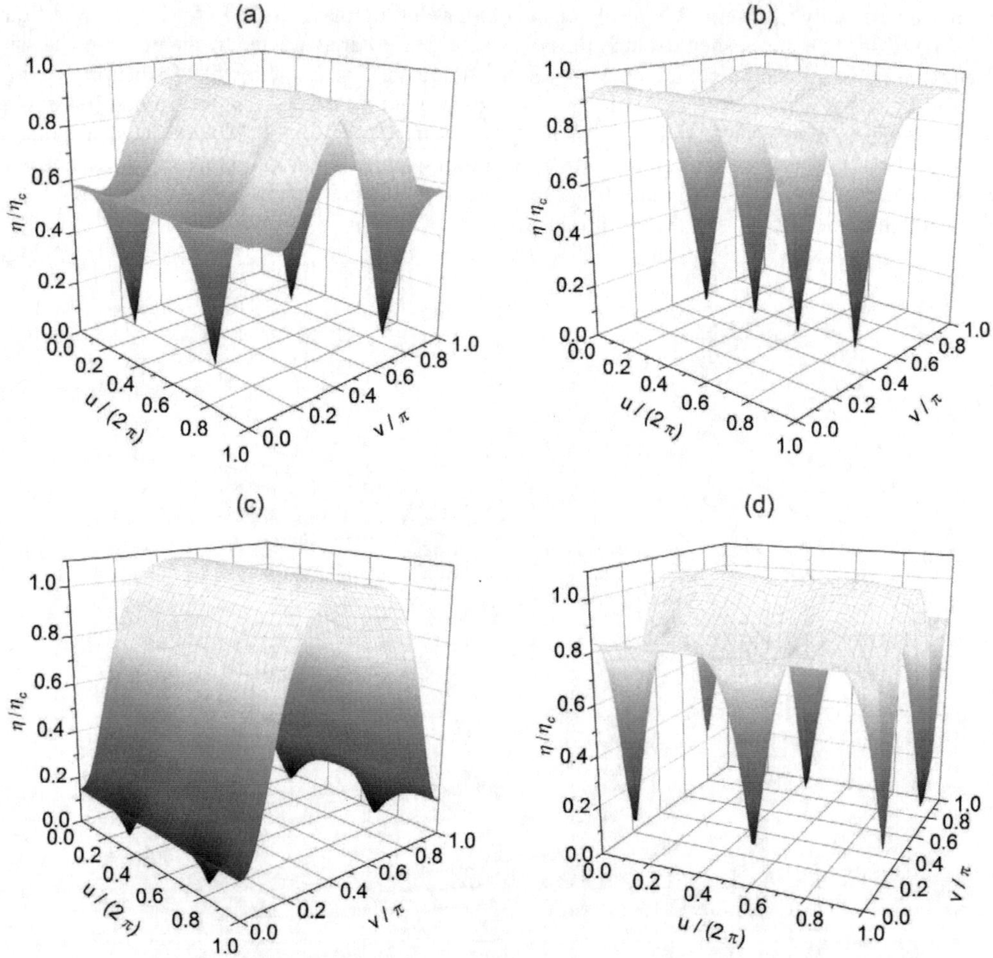

Figure 3. Representation in the $(\mathbf{u}, \mathbf{v})$-plane of the nematic textures presented in figure 2. Parameters are defined in the figure caption of figure 2.

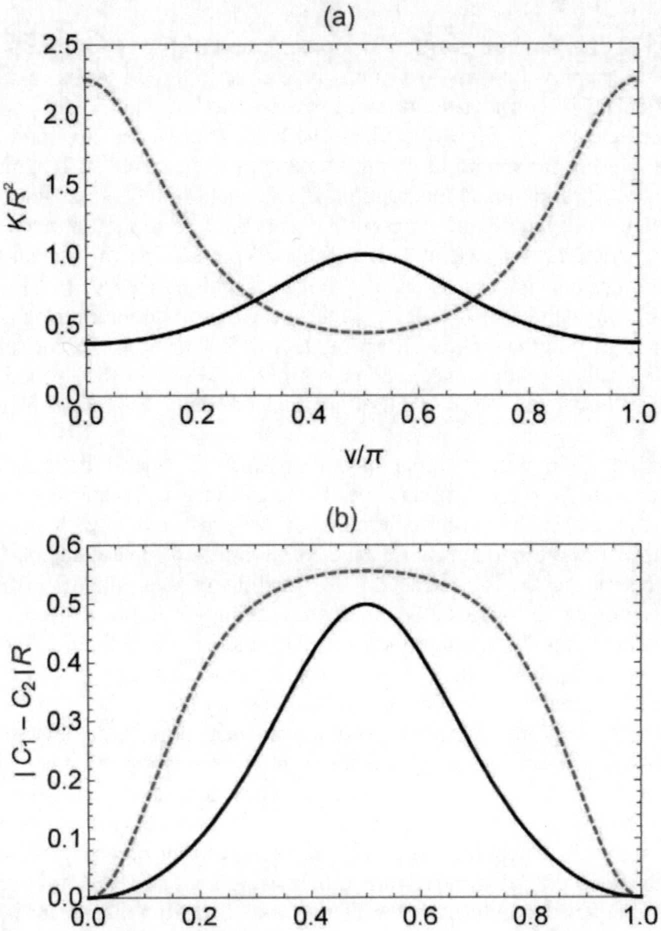

Figure 4. Gaussian curvature $\boldsymbol{K}$ (a) and the absolute difference between the principal curvatures $|\boldsymbol{C_1} - \boldsymbol{C_2}|$ (b) as the functions of the ellipsoid's zenith angle $\boldsymbol{v}$. Red dashed lines represent the prolate ellipsoid $\boldsymbol{a/b} = \mathbf{1.5}$, while the black lines represent the oblate ellipsoid $\boldsymbol{b/a} = \mathbf{1.28}$.

In prolate and oblate structures in general principal curvatures differ, i.e. $C_1 \neq C_2$. Consequently, for a finite value of k_e the *extrinsic* elasticity could play important role. For $k_e>0$ it tends to align nematic orientational ordering along principal directions exhibiting minimal curvature. This tendency is monotonously increasing with increasing difference $|C_1 - C_2|$, which is shown in figure 4b. Note that the resulting effective *extrinsic* field tends to align nematic ordering along its direction. Therefore, TDs tend to be expelled from regions where the effective *extrinsic* field is strong enough. In both prolate and oblate structures the difference $|C_1 - C_2|$ exhibits maximum in equatorial region. Consequently, in prolate structure TDs are pushed by the effective *extrinsic* field towards the poles, see figure 2c and figure 3c. In this case both *intrinsic* and *extrinsic* term are pushing TDs towards the poles. By contrast, in oblate structure the *intrinsic* and *extrinsic* term have competing tendencies. The *intrinsic* elasticity tends to assemble TDs in the equatorial region, while the *extrinsic* term expels them. In figures 2d and figure 3d we demonstrate cases where the effective *extrinsic* term is strong enough to expel TDs from the equatorial region.

4. Conclusions

We studied impact of curvature on position of topological defects in two-dimensional nematic films. For this purpose we employ Landau-de Gennes type description in terms of nematic tensor order parameter. In particular, we consider the role of *extrinsic* curvature [16,19,20], which has been so far in most studies neglected, see e.g. [13,14,15]. In the latter approaches covariant derivatives of order parameter were employed in expressing elastic free energy compatible with symmetry of systems of interest. Equivalently, it was assumed that minimal elastic distortion in a relevant order parameter field along a given direction obeyed parallel transport concept [21]. In a parallel transport along a line the order parameter experiences the least motion capable accompanying the spatial variation of surface normal. In the setting of our model $\vec{n}$ is parallel transported along a curve if it is seen as immobile in the frame $(\vec{e}_1, \vec{e}_2, \vec{v})$. In such cases it is known that the *intrinsic* curvature of a surface frustrates the order parameter field within the curved 2D space [13,14,15]. For example, for spherical topology the total integrated Gaussian curvature makes defects necessary even in the ground state. Furthermore, *intrinsic* curvature favours localization of TDs in regions of relatively high absolute value of Gaussian curvature.

We showed using a simplest possible modelling that in general both *intrinsic* and *extrinsic* curvature are expected to be present in elastic free energy density. Furthermore, there are no general reasons to claim that *extrinsic* curvature terms should be negligible with respect to the *intrinsic* contributions. The *extrinsic* term quantifies elastic costs of out-of-plane gradients in $\vec{n}$. Therefore, it is sensitive how 2D curved surface is embedded in 3D Euclidian space. The *extrinsic* coupling generates an unavoidable geometrically induced symmetry breaking field on surfaces patches possessing different principal curvatures. Its strength increases with increasing difference between the curvatures and equals zero in umbilic points, where the principal curvatures are equal. To illustrate this effect we calculated nematic structures on ellipsoidal shells. In the spherical geometry the *extrinsic* contribution is absent because all points are umbilics. However, on increasing prolatness or oblatness of shells regions where *extrinsic* curvature is present emerge. In our cases the effective *extrinsic* curvature is strongest at the equator of ellipsoids. Its ordering tendency is incompatible with TDs and it tends to expel them. This effect is most evident in oblate ellipsoids. Namely, the *intrinsic* curvature tends to localize TDs bearing positive topological charge in the equatorial region in order to screen strongly localised Gaussian curvature. However, for equally weighted *intrinsic* and *extrinsic* free energy contributions for strong enough oblateness the TDs are expelled from the equatorial region.

References

[1] Mermin N D 1979 *Rev. Mod. Phys.* **51** 591
[2] Zurek W H 1985 *Nature* **317** 505
[3] Lavrentovich O D 1998 *Liquid Crystals* **24** 117
[4] Hobson A 2013 *Amer. J. Phys.* **81** 211
[5] Skyrme T H R 1962 *Nucl. Phys.* **31** 556
[6] Rößler U K, Bogdanov A N and Pfleiderer C 2006 *Nature* **442** 797
[7] Ho T L 1998 *Phys. Rev. Lett.* **81** 742
[8] Meiboom S, Sethna J P, Anderson P W and Brinkman W F 1981 *Phys. Rev. Lett.* **46** 1216
[9] Kurik M V E and Lavrentovich O D 1988 *Physics-Uspekhi* **31** 196
[10] Vitelli V and Nelson D R 2006 *Phys. Rev. E* **74** 021711
[11] Gharbi M A, Seč D, Lopez-Leon T, Nobili M, Ravnik M, Žumer S and Blanc C 2013 *Soft Matter* **9** 6911
[12] Helfrich W 1973 *Z. Naturforsch. C* **28** 693
[13] MacKintosh F C and Lubensky T C 1991 *Phys. Rev. Lett.* **67** 1169
[14] Bowick M, Nelson D R and Travesset A 2004 *Phys. Rev. E* **69** 041102
[15] Vitelli V and Turner A M 2004 *Phys. Rev. Lett.* **93** 215301
[16] Selinger R L B, Konya A, Travesset A and Selinger J V 2011 *J. Phys. Chem. B* **115** 13989
[17] Kralj S, Rosso R and Virga E G 2011 *Soft Matter* **7** 670

[18] Mesarec L, Góźdź W, Iglič A and Kralj S 2016 *Sci. Reports* **6** 27117
[19] Napoli G and Vergori L 2012 *Phys. Rev. E* **85** 061701
[20] Napoli G and Vergori L 2012 *Phys. Rev. Lett.* **108** 207803
[21] Rosso R, Virga E G and Kralj S 2012 *Continuum Mech. Therm.* **24** 643
[22] Kamien R D 2002 *Rev. Modern Phys.* **74** 953